Saffron

Marketing and Value Chain Analysis

The Authors

Mr. Naveed Hamid possesses a meritorious academic record and awarded by university merit certificate in recognition of his having achieved the prescribed scholastic standards of high order OGPA in Masters (Agri Business Management) programme. He has obtained his degree in Agricultural Business Management from the Sher-e-Kashmir University of Agricultural Sciences and Technology of Jammu. He has 05 research publications, 01 abstract to his credit. He is member of two professional societies. He is awardee of prestigious project scheme awarded by National Bank for Agricultural and Rural Development in year 2014. His field of expertise is Agricultural Business Management and Marketing. Mr. Naveed Hamid is presently working as Assistant Professor (C) (Agri Business Management) in Division of Agricultural Economics & Agri Business Management, FoA, Chatha, Sher-e-Kashmir University of Agricultural Sciences and Technology of Jammu (J&K). In his two years experience of teaching and research , he has guided 2 research PG scholars.

Dr. Jyoti Kachroo has remained gold medalist (Two) and meritorious throughout her academic career, graduated and post graduate (Economics) from University of Jammu, Jammu and Kashmir. During her M.Phil and Ph.D. studies, she worked on Critical Appraisal on New Agricultural Strategy and Impact of New Agricultural Strategy on the economic and social constrictions of rural women respectively. She was initially appointed as Lecturer (Economics) in Govt. Degree Collage Jammu and later joined the Division of Agricultural Economics and Statistics, Sher-e-Kashmir University of Agricultural Sciences and Technology of Jammu as Associate Professor. Presently she is a Professor and Head, Division of Agricultural Economics and Agri business Management on SKUAST Jammu. Her field of expertise is women empowerment, agricultural production economics, marketing and econometric model based research. She has published about 80 articles which include book chapters, popular articles, review papers and about 32 research publications in referred journals. She has been honoured with D.T. Doshi Foundation award for best presentation in one of her research publications entitled "Economic Evaluation of Dry land and irrigated wheat based on stochastic model" by Agricultural Economics, Research Association New Delhi. She has 24 years experience of teaching and research and she has guided 16 research scholars.

Dr. Anil Bhat, Assistant Professor possesses meritorious academic record. He has done M.Sc. (Agril. Economics) from the Chaudhary Charan Singh University, Meerut and PhD (Agril. Economics) from the Sher-e-Kashmir University of Agricultural Sciences and Technology of Jammu. He also holds the Post Graduate Diploma in Rural Development from Inidra Gandhi National Open University, New Delhi and Post Graduate Diploma in Agricultural Marketing from Pondicherry University, Kalapet, Pondicherry. Dr. Bhat is presently working as Assistant Professor (Agricultural Economics) in Division of Agricultural Economics & Agri Business Management, FoA, Chatha, Sher-e-Kashmir University of Agricultural Sciences and Technology of Jammu (J&K). He has more than 25 publications to his credit. He is the member of four professional societies. His field of expertise is production marketing and agri business management.

Saffron

Marketing and Value Chain Analysis

Naveed Hamid
Jyoti Kachroo
Anil Bhat

2017
Scholars World
A Division of
Astral International Pvt. Ltd.
New Delhi – 110 002

Publisher's Note:

Every possible effort has been made to ensure that the information contained in this book is accurate at the time of going to press, and the publisher and author cannot accept responsibility for any errors or omissions, however caused. No responsibility for loss or damage occasioned to any person acting, or refraining from action, as a result of the material in this publication can be accepted by the editor, the publisher or the author. The Publisher is not associated with any product or vendor mentioned in the book. The contents of this work are intended to further general scientific research, understanding and discussion only. Readers should consult with a specialist where appropriate.

Every effort has been made to trace the owners of copyright material used in this book, if any. The author and the publisher will be grateful for any omission brought to their notice for acknowledgement in the future editions of the book.

Cataloging in Publication Data--DK
Courtesy: D.K. Agencies (P) Ltd. <docinfo@dkagencies.com>

Hamid, Naveed, author.
Saffron : marketing and value chain analysis / Naveed Hamid, Jyoti Kachroo, Anil Bhat.
pages cm
Includes bibliographical references.
ISBN 9789387057784 (International Edition)

1. Saffron (Spice)--India--Kashmir, Vale of--Marketing. 2. Value analysis (Cost control)--India--Kashmir, Vale of. I. Kachroo, Jyoti, author. II. Bhat, Anil (Assistant professor of agricultural economics), author. III. Title.

HF1054.I4H36 2017 DDC 338.17383 23

Published by : **Scholars World**
A Division of
Astral International Pvt. Ltd.
– ISO 9001:20150 Certi ied Company –
4736/23, Ansari Road, Darya Ganj
New Delhi-110 002
Ph. 011-4354 9197, 2327 8134
E-mail: info@astralint.com
Website: www.astralint.com

The Authors

Mr. Naveed Hamid possesses a meritorious academic record and awarded by university merit certificate in recognition of his having achieved the prescribed scholastic standards of high order OGPA in Masters (Agri Business Management) programme. He has obtained his degree in Agricultural Business Management from the Sher-e-Kashmir University of Agricultural Sciences and Technology of Jammu. He has 05 research publications, 01 abstract to his credit. He is member of two professional societies. He is awardee of prestigious project scheme awarded by National Bank for Agricultural and Rural Development in year 2014. His field of expertise is Agricultural Business Management and Marketing. Mr. Naveed Hamid is presently working as Assistant Professor (C) (Agri Business Management) in Division of Agricultural Economics & Agri Business Management, FoA, Chatha, Sher-e-Kashmir University of Agricultural Sciences and Technology of Jammu (J&K). In his two years experience of teaching and research , he has guided 2 research PG scholars.

Dr. Jyoti Kachroo has remained gold medalist (Two) and meritorious throughout her academic career, graduated and post graduate (Economics) from University of Jammu, Jammu and Kashmir. During her M.Phil and Ph.D. studies, she worked on Critical Appraisal on New Agricultural Strategy and Impact of New Agricultural Strategy on the economic and social constrictions of rural women respectively. She was initially appointed as Lecturer (Economics) in Govt. Degree Collage Jammu and later joined the Division of Agricultural Economics and Statistics, Sher-e-Kashmir University of Agricultural Sciences and Technology of Jammu as Associate Professor. Presently she is a Professor and Head, Division of Agricultural Economics and Agri business Management on SKUAST Jammu. Her field of expertise is women empowerment, agricultural production economics, marketing and econometric model based research. She has published about 80 articles which include book chapters, popular articles, review papers and about 32 research publications in referred journals. She has been honoured with D.T. Doshi Foundation award for best presentation in one of her research publications entitled "Economic Evaluation of Dry land and irrigated wheat based on stochastic model" by Agricultural Economics, Research Association New Delhi. She has 24 years experience of teaching and research and she has guided 16 research scholars.

Dr. Anil Bhat, Assistant Professor possesses meritorious academic record. He has done M.Sc. (Agril. Economics) from the Chaudhary Charan Singh University, Meerut and PhD (Agril. Economics) from the Sher-e-Kashmir University of Agricultural Sciences and Technology of Jammu. He also holds the Post Graduate Diploma in Rural Development from Inidra Gandhi National Open University, New Delhi and Post Graduate Diploma in Agricultural Marketing from Pondicherry University, Kalapet, Pondicherry. Dr. Bhat is presently working as Assistant Professor (Agricultural Economics) in Division of Agricultura Economics & Agri Business Management, FoA, Chatha, Sher-e-Kashmir University c Agricultural Sciences and Technology of Jammu (J&K). He has more than 25 publications his credit. He is the member of four professional societies. His field of expertise is producti marketing and agri business management.

Preface

With global demand for spices and spices products rising, India is targeting exports of 2.3 billion US dollars in 2013-14 fiscal even as Rs 9,433 crore worth spices were shipped during April-December last year. The total volume of spices and spices products exported in the nine month period of April-December 2013 was 5,71,680 tonnes, valued at Rs 9,433 crore, a 41 per cent growth in rupee terms and 27 per cent both in volume and dollar terms. During the same period of the previous fiscal, as much as 4, 49,926 tonnes valued at Rs 6,696 crore ($1232 million) was exported. For 2014-15 fiscal also the board was targeting spices exports of $2.3 billion. (Spice board of India). The countries like Iran, Spain and India are the major saffron producing nations of the world with Iran contributing to the 88 per cent of the world's saffron production and at the same time India contributes around 7 per cent of the total production with the average productivity of 2.30kg/ha (Anonymous). Production of saffron in India is restricted to the states of Jammu and Kashmir and Himachal Pradesh with the area of around 4265 hectares and annual production 7.50 MT. out of this about 2469.02 hectares lie exclusively in Jammu and Kashmir State

In Jammu and Kashmir which is considered as the second largest contributor of saffron to the global market, the cultivation is confined to district Pulwama, Budgam and Doda (Kishtwar). District Pulwama accounts for 75 per cent of total area under saffron in the state followed by the District Budgam accounting for 16.13 per cent of the total area. District Srinagar accounts for 6.68 per cent whereas, Poochal, Namil, Cherrad, Hullar, Blasia, Gatha, Bandakoota and Sangrambatta of District Kishtwar account for 2.5 per cent of the total area of the state. Due to the many uses of saffron crop its demand is growing faster than others in global as well as in domestic market. Various problems especially market related problems have emerged which needed careful investigation. Hence, it was found necessary to investigate the Value chain of of Saffron, price spread analysis, Problems faced by various stakeholders/chain holders and farmers in cultivation of Saffron crop in the district of Kashmir region which thereby can help the policy makers for its improvement in Production and Marketing. In this book, an attempt has been made to investigate "Value chain of saffron in Kashmir Valley of J&K State.

First chapter provides an overview on the status and importance of the saffron crop. Second chapter deals with the theory of value chain and marketing efficiency concept and conceptual frame work of the value chain and marketing efficiency. The literature regarding the financial assistance disbursement, government support, value chain aspect, marketing efficiency and price spread analysis aspect are detailed in third chapter. Fourth chapter deals with the results of various objectives and last chapter presents the summary and conclusion related to the study.

We felt our heartfelt gratitude to Dr. Dileep Kachroo (Registrar SKUAST-J) whose genuine interest, guidance and conceptual technical ideas during the course of planning and timely help in finalizing this book.

None is forgotten but everyone is not included.

Naveed Hamid

Jyoti Kachroo

Anil Bhat

Contents

List of Figures

List of Tables

Chapter 1

Saffron Production

An Overview

Saffron falls under the category of spices and the spice board of India (The Spices Board of India (Ministry of Commerce, Government of India) is the apex body for its export promotion. Established in 1987, the Board is the catalyst of these dramatic transitions. The Board plays a far reaching and influential role as a developmental, regulatory and promotional agency for Indian Spices. Spices are strongly flavoured or aromatic substance of vegetable origin, obtained from tropical plants, commonly used as a condiment". Spices were once as precious as gold. India plays a very important role in the spice market of the world. India produces a wide range of spices. Having varying climates from tropical to sub-tropical to temperate almost all spices grow splendidly in India. Almost all the states and union territories of India grow one or the other spices. Under the act of Parliament, a total of 52 spices are brought under the purview of Spices Board. However 109 spices are notified in the ISO list. The Indian spices can be categorized into three main categories as basic, complimentary and aromatic or secondary spices (Saffron). At present, production is around 3.2 million tonnes of different spices valued at approximately 4 billion US $, and holds a prominent position in world spice production. With global demand for spices and spices products rising, India is targeting exports of 2.3 billion US dollars in 2013-14 fiscal even as Rs 9,433 crore worth spices were shipped during April-December last year. The total volume of spices and spices products exported in the nine month period of April-December 2013 was 5,71,680 tonnes, valued at Rs 9,433 crore, a 41 per cent growth in rupee terms and 27 per cent both in volume and dollar terms. During the same period of the previous fiscal, as much as 4, 49,926 tonnes valued at Rs 6,696 crore ($1232 million) was exported. For 2014-15 fiscal also the board was targeting spices exports of $2.3 billion. **(Spice board of India).** Among the spices, Saffron (crocus sativus) is a perennial herb which belongs to the Iridaceae family and is the most expensive species in the world for its aroma and colour. It is the world's most favored species and its monopolistic character makes it more and more expensive. The three stigmas of the saffron flower are the most important economic part of the plant. The saffron stigma is rich in aroma and colour. In dried or powdered forms, stigmas are commonly used as "Spice or colouring in food preparation, Materials in pharmaceutical, cosmetic and perfume industries and Dye material in textile production.

The countries like Iran, Spain and India are the major saffron producing nations of the world with Iran contributing to the 88 per cent of the world's saffron production and at the same time India contributes around 7 per cent of the total production with the average productivity of 2.30kg/ha (Anonymous). Production of saffron in India is restricted to the states of Jammu and Kashmir and Himachal Pradesh with the area of around 4265 hectares and annual production 7.50 MT. out of this about 2469.02 hectares lie exclusively in Jammu and Kashmir State.

1.1 Saffron and its Uses

Saffron (Crocus sativa) which belongs to Iris family Iridaceae is the most expensive spice in the world and is popularly known as the "Golden Condiment". In India it is a legendary crop of Jammu and Kashmir, produced on well drained karewa soils where ideal climatic conditions are available for good shoot growth and flower production. Plants of this family are herbs with rhizomes, corms or bulbs. The family Iridaceae embraces about 60 genera and 1,500 species. The genus Crocus includes native species from Europe, North Africa and temperate Asia, and is especially well represented in arid countries of south-eastern Europe and Western and Central Asia. Among the 85 species belonging to the genus Crocus, C. sativus L. (Saffron) is the most fascinating and intriguing species (Fernåndez, 2004). Dried stigmas of saffron (Crocus sativus), a perennial herb, well-known for its aroma and used for flavouring, is a culinary delight. It is an important commodity and is of great significance in the agricultural economies of Iran (Khorasan) and India (Jammu and Kashmir). While saffron is well known as a spice, it has many other uses in industries such as food, pharmaceutical, cosmetic and perfumery as well as in the textile dyes (Kafi *et al.*, 2006; Mir, 1992).

The State of Jammu and Kashmir (Kashmir) is situated between 32° 17′ and 36° 58′ north latitude and 37°26′ and 80° 30′ east longitude and falls in the great north-western range of the Himalayas and constitutes the northern most extremity of India. Saffron (Crocus sativus Kashmirianus) covers about 4 per cent of the total cultivated area of the Kashmir valley and provides about 16 per cent of total agricultural income (Anonymous 2008). Saffron is chiefly grown in the following districts: Pulwama comprising Pampore, Balhuma, Wayun, Munpora, Mueej, Konibal, Dusoo, Zundhur, Letpora, Sombora, Barsoo, Ladu and Khrew; Budgam (16.13 per cent) comprising Chadura Nagam, Lasjan, Ompora and Kralpura; Srinagar comprising Zewan, Zawreh and Ganderbal; Doda comprising Poochal, Namil, Cherrad, Huller, Blasia, Gatha, Bandakoota and Sangrambatta, and some areas of Anantnag district comprising Zeripur, Srechan, Kaimouh, Samthan and Buch. Saffron cultivation forms an important sector for the livelihood security of more than 16,000 farm families located in 226 villages. The limited size of land holdings makes cultivation less profitable, with over 61 per cent of holdings below 0.5 ha, and only 26 per cent of holdings between 0.5-1.0 ha and 13 per cent of holdings > 1.0 ha (Anonymous 2009). The total area under this crop in the State of Jammu and Kashmir has shown a decrease of 83 per cent in the last decade, a 215 per cent decrease in production and a 72 per cent decrease in productivity.

1.2 Method of Planting

In Kashmir, at the advent of spring, *i.e.* in late March, which lasts till April, the field is repeatedly ploughed either by a plough or twice by tractor at an interval of about 20 days subsequently, in August leveling and hoeing operations are carried out. Fields are pulverized three to four times. Ideally 15-20 t of well-decomposed farm manure per hectare of land are applied and thoroughly mixed into the soil before the last tillage operation (Mir, 1992; Munshi *et al.*, 2001; Zargar, 2001). The field is laid out into 1.5-2 meter wide strips (normally 2-3 m long) across the field with slopes along the sides, which end up in 30 cm wide 20 cm deep drainage channels on both sides. These channels help in draining out the excess water. The corms are sown 12-15cm below these raised beds at a distance of 10x20 cms. The most suitable time for sowing the corms is from the last week of August till mid September, however, some farmers also sow corms after this period, but the yield is not up to the expectations (Mir, 1992; Munshi *et al.*, 2001; Zargar, 2001). In Khorasan is saffron planted either in dry or moist beds. In traditional systems, corms are planted in hills 25 cm apart and with sometimes up to 15 corms per hill with no row arrangement. When corms are planted in rows, shallow ditches 30-35 cm apart is made by a furrower and corms are arranged in hills of 3 to 15 corms and finally covered with soil. Flat bed planting in Khorasan has been reported to be advantageous compared with furrow planting. Since saffron is solely reproduced by corms, its size and health is crucial for productive farming. Small corms do not have the potential to produce flowers in the first year (Sadeghi, 1983). Corms weighing more than 5g are capable of producing flowers in the first year. In Khorasan corms with a weight of more than 5 g are selected for planting. Farmers are not particular about the selection of the corm size in Kashmir. They sow 3-15 corms in each hill to compensate for the number of flowers produced by smaller corms. In one investigation in Shiraz, Iran, it was shown that corms weighing less than 8g were not recommended (Sadeghi, 1983; Shirmohamadi and AliakbarKhani, 2002).

In Kashmir, the size of the corm is usually expressed by its diameter and corms diameter above 2 cm are recommended for sowing. It is not common in Kashmir to sow more than one corm in each hole (Hassan *et al.*, 2001; Munshi *et al.*, 2001). Therefore, the number of corms sown in one unit of land is much lower than that of Khorasan. For instance, in one m2 of saffron farm in Khorasan 150-250 corms are sown, if the average corm size be 5 g, more than 10 t/ha of corm are needed. Kashmiri farmers sow less than half of this amount; about 5 t of corms are sown per hectare of land. This could be the main reason why the farmers of Kashmir harvest negligible saffron flowers in the first year and in some areas in the second year after sowing.

1.3 Saffron Usage

Saffron contains many plants derived chemical compounds that are known to have been anti-oxidant, disease preventing and health promoting properties. The flower stigma are composed of many essential volatile oils but the most important being safranal, which gives saffron its distinct hay-like flavor. Other volatile oils in saffron are cineole, phenethenol, pinene, borneol, geraniol, limonene, p-cymene, linalool, terpinen-4-oil, etc. This colorful spice has many non-volatile active components; the most important of them is α-crocin, a carotenoid compound, which gives the stigmas their characteristic golden-yellow color. It also contains other carotenoids, including zea-xanthin, lycopene, α- and β-carotenes. These are important antioxidants that help protect the human body from oxidant-induced stress, cancers, infections and acts as immune modulators. The active components in saffron have many therapeutic applications in many traditional medicines as antiseptic, antidepressant, anti-oxidant, digestive. This novel spice is a good source of minerals like copper, potassium, calcium, manganese, iron, selenium, zinc and magnesium. Potassium is an important component of cell and body fluids that helps control heart rate and blood pressure. Manganese and copper are used by the body as co-factors for the antioxidant enzyme, superoxide dismutase. Iron is essential for red blood cell production and as a co-factor for cyto chrome oxidizes enzymes.

Additionally, it is also rich in many vital vitamins, including vitamin A, folic acid, riboflavin, niacin, vitamin-C that is essential for optimum health.

1.4 Medicinal Uses

The active components present in saffron have many therapeutic applications in many traditional medicines since long time ago as anti-spasmodic, carminative, diaphoretic. Research studies have shown that, safranal, a volatile oil found in the spice, has antioxidant, cytotoxicity towards cancer cells, anticonvulsant and antidepressant properties. Alfa-crocin, a carotenoid compound, which gives the spice its characteristic golden-yellow color, has been anti-oxidant, anti-depressant, and anti-cancer properties. It acts as Flavoring digestion and strengthens the function of stomach, sedative which combats cough and Bronchitis, Mitigates colic and insomnia, Calming effect on infants during teething fits, Flavoring expulsion of gases accumulated in digestive track, an anti spasmodic, Excellent against headaches, when applied as a paste to a forehead, and as an antidepressant.

Saffron has an aroma and flavor which cannot be duplicated, and a chemical makeup which, which when understood, helps the chef or home cook to know how to best release the flavor and aroma in cooking and baking. Saffron is sold in two forms, powder and threads and each behave very differently in the kitchen. Saffron has a strong perfume and bitter, honey like flavor. The taste is pleasantly spicy and bitter and the odor is tenacious. It has strong perfume and bitter honey taste. Saffron is mainly used as a colorant and flavor for cheeses, pastry, and rice and sea food dishes. Saffron is used in spice blends for paella, curry kheer and bouillabaisse. Saffron is native to the Mediterranean and is grown in Spain, France, Portugal, India and Italy. Spain is considered the premier source of saffron. Its flavor is distinctive and agreeable in character.It is native to Asia Minor, where it has been cultivated of thousands of years to be used in medicines, perfumes, dyes and a wonderful flavoring for foods and beverages. Saffron is nowadays used in making a sizzling cup of saffron drink, as it adds a value to the simple drink by its coloring and fragrant attribute.

***Source*: Google.**

1.5 World Saffron Production

Saffron is currently being cultivated more or less intensely in Iran, India, Greece, Morocco, Spain, Italy, Turkey, France, Switzerland, Israel, Pakistan, Azerbaijan, China, Egypt, United Arab Emirates, Japan and recently in Australia (Tasmania). The world's total annual saffron production is estimated at 205 tons per year. Iran is said to produce 80 percent of this total; *i.e.*, 160 tons and its Khorasan province alone 137 tons of the totals. The age of saffron farms there is the most important factor influencing yield and five years aged farms had the longest flowering period, also there is a positive linear relation between continuance of flowering and yield. The Kashmir region in India produces between 8 to 10 tons mostly dedicated to India's self-consumption. Greek production is 4 to 6 tons per year. Morocco produces between 0.8 and 1 ton. Saffron production has decreased rapidly in many traditionally producing countries and is abandoned in others such as England and Germany. Spain used to be the traditional world leader and most reputed saffron producer for centuries in areas of La Mancha and Teruel. Nowadays, the production is only about 0.3-0.5 tons. Production of Italy (Sardinia, Aquila, Cascia) 100 kg;

Turkey (Safranbolu) 10 kg; France (Gâtinais, Quercy) 4-5 kg; and Switzerland (Mund) 1 kg are nearly insignificant. All saffron producers in the European Union, as well as in Turkey, suffer from increasing labour costs. However, the demands of world industry for saffron product are very high. Among countries cultivating saffron, Iran is at the first place. More than 90 per cent of the world production falls onto Iranian saffron which has a great importance in the economy of the country. The area under saffron cultivation in Iran reaches about 80,000 hectares with an annual production of about 250 tons of the dry stigmas. In recent years this amount has been remarkably increased which was achieved mainly by extension of the planting areas but not due to the increased yield capacity in the area unit. In Kashmir, Beside other factors new high yielding cultivars of saffron are required to solve the problem.

Table 1.1: Country-Wise Area, Production and Productivity (2014)

Country	*Area (ha)*	*Production (MT)*	*Yield (kg/ha)*
Iran	43,408 (87.7)	174.00 (88.89)	4.00
India	4265 (6.59)	7,50 (3.83)	2.29
Greece	1000 (2.02)	4.30 (2.19)	4.30
Azerbaijan	675 (1.36)	3.70 (1.88)	5.48
Spain	600 (1.21)	5,00 (2.54)	2.00
Morocco	500 (1.01)	1,00 (.50)	2.00
Italy	29.4 (.06)	0.24 (0.12)	8.16
Total	49477.4	195.74	3.96

Source: Annual Report of Directorate of Agricultural Jammu and Kashmir.

Note: Figures in parentheses are percentages to total.

The leading saffron producing countries of the world like Iran, Spain and Greece with intensive production technologies are able to achieve higher production and productivity 4-8kg/ha which is much higher than productivity in Kashmir, as a result it has poses a threat to the saffron industry due to increasing imports of saffron every year. The data has revealed that during the year 2006-07 total saffron imported from different countries was 3.3 metric tons and it was 3.7 tonnes in 2010-11. With the prevailing climatic conditions causing uncertainty such as drought conditions what needs to be stressed is that saffron here is a rain fed crop, the soils in Kashmir require irrigation and are comparatively less productive, the soils are also overloaded with pathogenic fungi and rodents. Thus there is a need to bring more and more area under cultivation (infact restore to the earlier acreage levels) and increase the average productivity by adopting intensive production system. The government has certainly initiated progressive steps in enhancing the saffron production in the valley launching projects like "value chain on Kashmiri saffron" with the support of Indian Council of Agricultural Research using environmental friendly techniques and the "National Saffron Mission" through the investment of Rs 3760 million, but a major mutually coordinated push by the agencies involved is required to increase production and productivity levels.

1.6 Saffron Production in Jammu and Kashmir

In Jammu and Kashmir which is considered as the second largest contributor of saffron to the global market, the cultivation is confined to district Pulwama, Budgam and Doda (kishtwar). District Pulwama accounts for 75 per cent of total area under saffron in the state followed by the District Budgam accounting for 16.13 per cent of the total area. District Srinagar accounts for 6.68 per cent whereas, Poochal, Namil, Cherrad, Hullar, Blasia, Gatha, Bandakoota and Sangrambatta of District kishtwar account for 2.5 per cent of the total area of the state. As per the official sources in the past 10 years there has been a steady decline in saffron production due to shift of agricultural land to the commercial purposes, in 1998 the crop was grown over 4161 hectares which has come down to 2469.02 hectares in 2011-12 which is the threat to the saffron industry in J&K. However state and central government and some research intuitions have taken some important steps to boost the value of Kashmiri saffron with the world bank launching a project called "value chain on Kashmiri saffron" which aim to boost production and quality using environment friendly techniques, and there is being demonstration of plots in J&K and national saffron mission to promote the saffron production in Jammu and Kashmir. Apart from this government has promised of providing 70 Kanals of land for setting up the saffron research centre to fetch a good amount on the account of exports. Low prices and competition from cheaper imported Iranian saffron have forced many

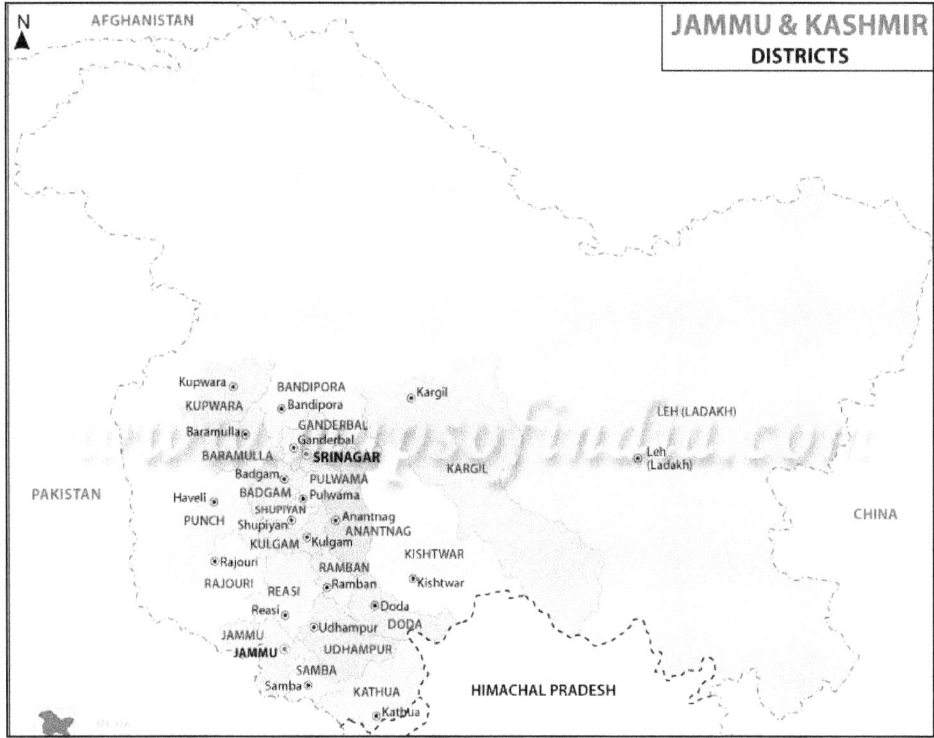

Map of Jammu and Kashmir Region.

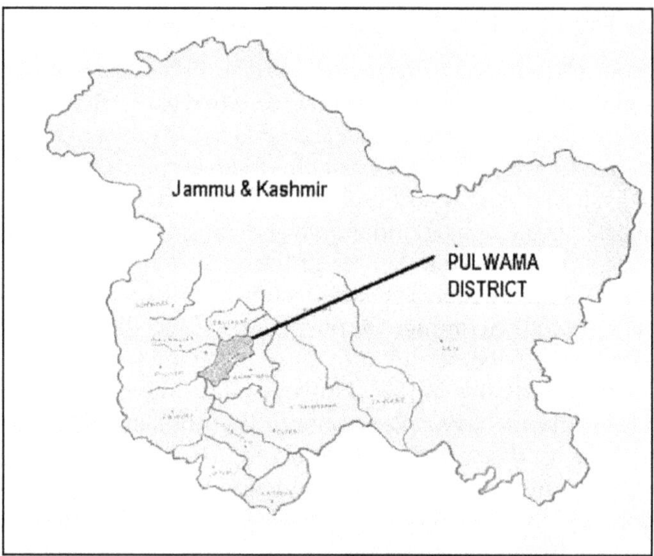

Map of Saffron Area.

traditional saffron farmers to abandon this crop so government need to step in urgently to protect this sector before it is too late. With the passage of time the saffron of Kashmir is dying that can be guarded/noticed from the fact that its production has decreased from 15.59MT to 10MT.

Chapter 2

Theory of Value Chain and Marketing Efficiency

Value chain is the sequential set of primary and support activities that an enterprise performs to turn inputs into value-added outputs for its external customers. As developed by Michael E. Porter, it is a connected series of organizations, resources, and knowledge streams involved in the creation and delivery of value to end customers. Value systems integrate supply chain activities, from determination of customer needs through product/service development, production/operations and distribution, including (as appropriate) first-, second-, and third-tier suppliers. The objective of value systems is to position organizations in the supply chain to achieve the highest levels of customer satisfaction and value while effectively exploiting the competencies of all organizations in the supply chain. The idea of the value chain is based on the process view of organizations, the idea of seeing a manufacturing (or service) organization as a system, made up of subsystems each with inputs, transformation processes and outputs. Inputs, transformation processes, and outputs involve the acquisition and consumption of resources - money, labour, materials, equipment, buildings, land, administration and management. How value chain activities are carried out determines costs and affects profits.

2.1 Concept of Value Chain

Michael E. Porter was the first to introduce the concept of a value chain. Porter, who also developed the Five Forces Model that many businesses and companies use to figure out how well they can compete in the current market place, first discussed the value chain concept in his book "Competitive Advantage: Creating and Sustaining Superior Performance" (Free Press, 1985)."Competitive advantage cannot be understood by looking at a firm as a whole," Porter wrote. "It stems from the many discrete activities a firm performs in designing, producing, marketing, delivering and supporting its product. Each of these activities can contribute to a firm's relative cost position and create a basis for differentiation. Porter suggests that activities within an organization add value to the service and products that the company produces, and that all of these activities should be run at optimum level if the organization is to gain any real competitive advantage. If they are run efficiently, the value obtained should exceed the costs of running them. *E.g.* customers should return to the company and transact freely and willingly.

In his book, Porter said a business's activities could be split into two categories: primary activities and support activities. Primary activities include the following:

2.1.1 **Inbound logistics:** This refers to everything involved in receiving, storing and distributing the raw materials used in the production process.

2.1.2 **Operations:** This is the stage where raw products are turned into the final product.

2.1.3 **Outbound logistics:** This is the distribution of the final product to consumers.

2.1.4 **Marketing and sales:** This stage involves activities like advertising, promotions, sales-force organization, selecting distribution channels, pricing, and managing customer relationships of the final product to ensure it is targeted to the correct consumer groups.

2.1.5 **Service:** This refers to the activities that are needed to maintain the product's performance after it has been produced. This stage includes things like installation, training, maintenance, repair, warranty and after-sales services.

The support activities help the primary functions and comprise the following:

2.1.6 **Procurement:** This is how the raw materials for the product are obtained.

2.1.7 **Technology development:** Technology can be used across the board in the development of a product, including in the research and development stage, in how new products are developed and designed, and process automation.

2.1.8 **Human resource Management:** These are the activities involved in hiring and retaining the proper employees to help design, build and market the product.

2.1.9 **Firm infrastructure:** This refers to an organization's structure and its management, planning, accounting, finance and quality-control mechanisms.

2.2 Identify and Understand your Company's Value Chain

For each primary activity, determine which specific sub activities create value. There are three different types of sub activities:

2.2.1 Identify Sub Activities for each Primary Activity

Direct activities create value by themselves. For example, in a book publisher's marketing and sales activity, direct sub activities include making sales calls to bookstores, advertising, and selling online.

Indirect activities allow direct activities to run smoothly. For the book publisher's sales and marketing activity, indirect sub activities include managing the sales force and keeping customer records.

Quality assurance activities ensure that direct and indirect activities meet the necessary standards. For the book publisher's sales and marketing activity, this might include proofreading and editing advertisements.

2.2.2 Identify Sub Activities for each Support Activity

For each of the Human Resource Management, Technology Development and Procurement support activities, determine the subactivities that create value within each primary activity. For example, consider how human resource management adds value to inbound logistics, operations, outbound logistics, and so on. As in Step 1, look for direct, indirect, and quality assurance sub activities.

Then identify the various value-creating subactivities in your company's infrastructure. These will generally be cross-functional in nature, rather than specific to each primary activity. Again, look for direct, indirect, and quality assurance activities.

2.2.3 Identify Links

Find the connections between all of the value activities you've identified. This will take time, but the links are key to increasing competitive advantage from the value chain framework. For example, there's a link between developing the sales force (an HR investment) and sales volumes. There's another link between order turnaround times, and service phone calls from frustrated customers waiting for deliveries.

2.3 Marketing Efficiency

Marketing efficiency is considered to be a pre-requisite for prompt delivery of goods. Prompt delivery of good at a reasonable price is possible only if the market works in a competitive way. Competitive mechanism is possible only when the market agents are free to exercise their actions. An efficient marketing system implies that price spread or marketing margin is fairly less. In market integration terminology, prices in spatially separated markets will get differed only by transaction costs among markets. Lower price spread also implies that both consumers and producers are gaining from affordable price and reasonable profit. Hence, an efficient marketing system implies the existence of market integration.

2.4 Review of Studies Related to the Value Chain Process

Bhagat (1982) in his study on "Growth Rates of Output of Coarse Grains and Pulses in a Backward Economy: A Study of Temporal-Spatial Variations in

Chotanagpur", examined the magnitude and direction of changes in area, yield rate and output of major coarse food grains and pulses in relation to their competing crops in the districts of Chotanagpur region in Bihar in the two sub-periods, *i.e.*, pre-high Yielding Varieties (HYV) (1956-57 to 1965-66) and post-HYV (1967-68 to 1976-77) period along with the causes underlying these changes and variations. It was found that there was no marked difference in the growth rates of crops grown during the rabi season in the pre-HYV period but the growth rates of area and output of barley and gram were found to be much lower than that of wheat in the HYV period, indicating the farmers" preference for growing wheat as against barley and gram, however the reasons for comparatively higher growth of wheat area and output were not investigated.

Singh (1981) while conducting study on "Imbalances in Agricultural Growth" and found that an uneven growth rate of individual crop has led to the regional imbalances in the rural prosperity. Rice is the main cereal crop in most of states, even in this crop also the growth rate has not been uniform in the different states. Agriculture is the main income generation factor in rural India. There is any imbalance evidence in growth led to imbalance among the regions and states.

Rath (1980) examined the performance of agricultural production in India. An exponential trend function was used to estimate production, are and yield rates for the period from 1949-50 to 1977-78. The analysis indicated that the total agricultural production of India grew at an average rate of 2.38 per cent per year during the years 1955-56 to 1978-79. The rate of growth was found to be somewhat higher during the 10 years ending 1964-65 than during the subsequent 13 years when it was only about 2.42 per cent. It was also observed that the sustained growth of food grains was entirely due to cereals. Wheat recorded a growth rate of 3.9 per cent and rice of about 3 per cent. In the post-1965 period, with the advent of new High Yielding Varieties (HYVs), the growth rate of wheat almost doubled to 7.06 per cent of this about half was due to increase in area and other half due to increase in productivity of the crop. However, wheat production which increased at the rate of about 11 per cent during 1964-72 registered only 3 per cent growth rate during seventies. But this was not surprising as high yielding wheat had already reached very large areas and achieved economically high level of yield.

Dantwala (1978) presented a brief review of the anatomy of agricultural growth in India and reported that the extent of irrigation is the main factor which explains growth rate in the country. With regard to rice farming, about 77 million hectares (53 per cent) of world rice area is irrigated. 70 to 75 per cent of world rice production also comes from irrigated areas.

Grewal and Rangi (1983) made an attempt to estimate the growth of Punjab agriculture in relation to productivity increase and response to the use of fertilizer and observed that productivity increase was largely the outcome of increased irrigation and fertilizer use. It was further estimated that one kilogram of fertilizer nutrients yielded about 8 kg of wheat grains during the period 1967-68 to 1976-77 when the average use of fertilizer for wheat crop was about one-third of the recommended dose. During the succeeding period ending 1981-82, this response was only 5.6 kg of wheat per kg of fertilizer nutrients applied. This was the case when fertilizer use was two-third of the recommended level.

Ray (1983) analyzed the sources of change in the pattern of growth and instability in Indian agriculture for the period 1950-80. Rice, wheat, coarse cereals, cereals, pulses, foodgrains, oilseeds, sugarcane, cotton, jute and tobacco crops were chosen for the study. The study reveals that instability in production was lower during the 1950s for all the crops and crop groups except tobacco. 1960s and 1970s recorded higher instability in production of all crops which can be attributable to new agricultural technology. Pulses suffered from more severe amplitude of fluctuations. The study concludes that with rapid growth, stability can be achieved if the environment for production is brought under human control and even with a slower growth, production can be made more unstable through price polices.

Bandyopadhyay (1989) while conducting a study on "Growth and Instability in the Production of Main Cereal Crops in West Bengal and Punjab-Haryana", reported that the elaborate network of irrigation in Punjab-Haryana, independence of the vagaries of the monsoon/rainfall along with the development of private tube well irrigation on a substantial scale, particularly, during post-green revolution period, may have effectively narrowed down the year-to-year fluctuations in the production of wheat. Rice crop has also been introduced in the states which grew at a much faster rate compared to that in West Bengal. Irreparable damage done to the ancient irrigation network in Bengal during the colonial rule on the other hand, has not been compensated by the development of irrigation during the era of planned development since independence in West Bengal. Irrigation has been the main technological constraint for agricultural production in West Bengal.

Ray (1991) studied the instability in Indian agriculture and found that the turning point, both in rice and wheat production growth rate, appeared to have occurred in the early 1970s. The increased use of HYVs of seeds and modern purchased inputs for accelerating crop output growth rate have made crop production more sensitive to weather and price changes. On the basis of empirical results, it has been concluded that the production instability is and inevitable consequence of rapid agricultural growth and there is little that can effectively be done about it. The contrasting pictures obtained from rice and wheat production data illustrate the point. Even with a slower growth rate, production can be made more unstable by stability is caused by uncontrollable natural factors; its intensity can be progressively reduced by effective to reduce the instability of cereal production in Haryana.

Jha (1994) while conducting study on growth and instability in agriculture associated with new agricultural technology during 1972-73 to 1990-91, found decline in yield instability in gross return and yield over years. The decline in yield instability in crops, *viz.*, paddy and wheat was brought about with increased area under irrigation over the years. Nevertheless, the government's consistent price policy also helped in reduction of instability in farm harvest prices. Thus, it has been inferred that with new technology, instability in agricultural income reduced with adequate irrigation facilities and consistent price policy. However, the hypothesis of high instability in agriculture accompanied with high growth rate was established.

Damodar (1996) found that the traditional agricultural of the district was capable of dynamic growth impulses in respect of the traditional crops. He found that there is no clear evidence for intensified adoption of HYV seeds in the selected crops and also for the impact of HYV seeds on the productivity of selected crops in Mahaboobnagar District.

Pochanna (2000) has observed that the association between yield growth and instability shows a negative direction at national and in most of states. He observed the positive association between yield growth and instability in Punjab, Haryana, Andhra Pradesh, Karnataka, Orissa and West Bengal. He found that the use of Fertilizers and spread of irrigation facilities are important factors not only to yield growth but also to reduce the fluctuations and ensure stability in the yield growth.

Rahane *et al.* (2000) examined in their article titled, "Trends in Area, Production and Productivity of Important Fruits and Vegetables in Maharashtra" the performance of fruits and vegetables in Maharashtra. The study was divided into two parts: (i) time series data on area, production and productivity of surveyed crops like coconut, cashew-nut, orange, grape, banana and onion during the period 1983-84 to 1997-98; and (ii) data on non-surveyed crops such as fruits (pomegranate, papaya, custard apple, etc.) and vegetables (chilly, garlic, potato, cauliflower, tomato, pea, etc.) during the period 1994-95 to 1997- 98. Annual growth rates were worked out to examine the region-wise performance of major fruits and vegetables grown in Maharashtra. The study brought out that in the case of surveyed crops and non-surveyed fruit crops, in spite of significant increase in area and production, their productivity decreased but in the case of non-surveyed vegetables, productivity remained constant in spite of increase or decrease in area.

Dahiya *et al.* (2001) examined the growth rates of area, production and yield of fruit crops in Haryana for the period 1990-91 to 1997-98. The results indicated that the area and production of guava and ber showed statistically significant increasing trend. In the case of mango, citrus and grapes, the area showed the significant increasing trend but production was found to be statistically non-significant. In the case of yield, all crops showed non-significant declining trend. The annual growth rate for fruit crops under the state registered a positive growth.

Shaheen and Shiyani (2004) in their paper entitled, "Growth and Instability in Area, Production and Yield of Fruit Crops in Jammu and Kashmir-A Disaggregate Analysis" concluded that the temporal change in area under different crops revealed a significant increase in area under apple, cherry and walnut over the time in J&K. Moderate to high significant growth was observed in area, production and yield of all the fruit crops for the period 1974-02 at the state level. The results of instability index indicated moderate to high instability in production and yield for all the fruits, except apple, which showed low instability for all the three parameters (area, production and yield) throughout the period. Higher instability in production in the case of perennial fruit crops was generally the consequent of instability in the productivity of the crop. The authors suggested to conduct an in-depth study in order to identify the factors responsible for higher instability in production and yield of crops like pear, apricot, peach, plum and cherry.

Kumar *et al.* (2004) has revealed that the Total Factor Productivity Growth (TFPG) of the crop sector in the Indo-Gengetic Plains by IGP had risen at the rate of 1.2 per cent per annum during the period 1980-81 to 1996-97. The TFP results for different agro-eco-regions have shown considerable variations. The Low-Gangetic Plain (LGP) region has depicted the highest growth in TFP (3.1 per cent) and Middle Gangetic (MGP), the lowest (0.37 per cent). The TFP growth rates were estimated at 1.4 per cent in the Trans-Gangetic Plain (TGP) and 0.9 per cent in the Upper-Gangetic Plains (UGP). In IGP, one-third of output growth was contributed by TFP. However, the contribution of TFP to output growth varied from as high as 57 per cent in the LGP to a meagre 17.3 per cent in the MGP. The shares of TFP in the output growth of the crop sector in the TGP and the UGP regions were observed to be 34 per cent and 26 per cent, respectively. The output growth in the UGP and the MGP was input-based, while in the LGP, it was technology-based. The output growth in the TGP was input- as well as technology-based. The analysis has confirmed that contribution of TFPG to output growth had started declining and was, in fact, showing a tendency of further deterioration in the process. Productivity growth, which picked up during the early-1980s, could not sustain during 1990s and this situation raised an alarm for the policymakers and researchers of the country.

Kumar and Mittal (2006) in their scholarly article on "Agricultural Productivity Trends in India: Sustainability Issues" discussed the sustainability issue of the crop productivity is fast emerging. The post-Green Revolution phase is characterized by high input-use and decelerating total factor productivity growth (TFPG). The agricultural productivity attained during the 1980s has not been sustained during the 1990s and has posed a challenge for the researchers to shift the production function upward by improving the technology index. It calls for an examination of issues related to the trends in the agricultural productivity, particularly with reference to individual crops grown in the major states of India. Temporal and spatial variations of TFPG for major crops of India have also been examined.

Chowdary (2011) in his paper article on "Agricultural Development in Assam" estimates that in Assam rice has registered a record production in 2010-11 at 50.86 lakh tonnes which marks a 15.4 per cent growth over the previous year and which accounts for more than 5 per cent of the country's total rice production. But, he regrets that basically an agro-based economy, Assam, the largest state of north-eastern region has remained poor because of agricultural backwardness, in the main. The very fact that Assam has to feed 2.6 per cent of India's population with 1.4 per cent of national income shows what development distance the state has to cover to catch up with the rest of the country. Even the other seven states of north eastern region stand on better footing since they together earn 1.3 per cent of national income to feed 1.2 per cent of the country's population. The per capita income distance of Assam from all India per capita at current prices registered an increase from Rs.757 in 1989-99 to Rs.3599 in 1990-2000 and onward to Rs.20, 250 in 2009-10.

Radhika *et al.* (2011) in their essay on "Climate Change and Rainfed Agriculture: Rural Development Perspectives" observed that erratic monsoon precipitation would adversely affect the lives of majority of population of this region.With the productivity of irrigated lands reaching a plateau, maintaining the food grain production of pace with the increasing population is a real challenge. However, in the present context, the scope for horizontal expansion of agriculture is almost nil as it is already extended to marginal lands. Protagonists of climate change argue that promotion of conventional agriculture of augment the vertical expansion will further deteriorate climate in the form of increase in global emissions through various activities of agriculture which is already held responsible for 18 per cent of global gas emissions. Hence, only vertical expansion that has to come mainly from rainfed region with sustainable and eco-friendly agriculture practices which are seen as sink for GHG emissions is the option left to make the production viable and to meet the food security in long run. Therefore, the paper attempts to understand the impact of climate change on agriculture, with particular reference to rainfed agriculture, present policy scenario and the strategies to be adopted with in-built climate change mitigation measures, in the context of rural development.

Dwivedi *et al.* (2011) in their scholarly article on "An Analytical Study of Capital Formation in India: With Special Reference to Indian Agriculture" says that an increase in the stock of capital in a fixed period is known as capital formation of an economy. The capital formation depends upon three factors- formation of savings, mobilization of savings and investment. Agriculture sector still dominates the Indian economic scene by providing livelihood to majority of the population. In most of the developing countries including India, agricultural development is a precondition for economic development. Capital formation is one of the basic factors for increasing production. This is all the more important in the agricultural sector where we are faced with the task of increasing production to keep pace with the increase in population against the odd of the vagaries of monsoon. The judicious use of natural resources for sustainable production of agriculture, adoption of advanced technology and development of infrastructure for facilitating all agriculture activities, ensuring food security in the broader sense of making adequate nutritious food available and accessible to all and making agriculture a profitable activity at par with other industries in the sphere of global economy are the problems that can be successfully tackled only with a strong capital base. This requires a close monitoring of the status of capital formation which is turn things on the nature of statistical system and quality of data available for measurement of capital formation. The share of agriculture in GDP has registered a steady decline from 18.6 per cent in 2004-05 14.6 per cent in 2009-10 at 2004-05 prices). The declining share of agricultural sector in the GDP is a characteristic of all developing economics. The Gross Capital Formation (GCF) in agriculture sector relative to GDP in this sector has an increasing trend from 15.8 per cent in 2005-06 to 21.3 per cent in 2008-09.

Kannan, (2011) analysed the long term growth patterns of major crops across the states, the compound annual growth rates calculated for the period 1967-68 to 2007-08 have been grouped into four categories, *viz.*, high (>4.0 per cent), medium (2.0-3.9 per cent), low (0-1.9 per cent) and negative. This long-term growth analysis will help to identify lagging crops and states and suggest suitable technology, policy and institutional options for accelerating growth. It showed growth in area of major crops by states. It is clear that a few oilseed crops like sunflower, rapeseed and mustard, sesamum and coconut registered more than 4.0 per cent growth in area in different states. Onion registered high growth rate in Gujarat, Karnataka and Maharashtra. Punjab was the only state, which showed high growth in area under rice. Similarly, potato emerged to be the major crop along with rapeseed and mustard, and sesamum in West Bengal. Surprisingly, coconut registered high growth in Tamil Nadu. At all-India level, only sunflower showed more than 4.0 per cent growth in area during 1967-68 to 2007-08.

Kaloo *et al.* (2011) analysed that the contribution of different variables influencing saffron production indicated that the regression coefficient of time (0.96) fertilizers (0.23) and area (0.13) contributed significantly to 0.48 per cent. Also auto correlation model was used: indicated (2.246) presence of perfect positive autocorrelation between the variables. Estimated calculation of area was calculated through the compound growth rate resulting 2469.023 ha will be used in 2012 for the cultivation purposes in 2012. Data used has shown a declining trend in the production and productivity from the last decade due to conversion of saffron land for commercial purposes.

Tantry (2014) evaluated the growth in terms of production, area and productivity trends of saffron production in Jammu and Kashmir. The study concluded that area, production and productivity of the precious and high value agricultural crop saffron in the state had depicted a declining trend since 1996-97. Area has declined from 5700 hectares in 1996-97 to 3000 hectares in 2007-08 and productivity decline from 3.72 kg/hectare in 2003-2004 to 2.52 kg/hectare. The study further concluded that the growth rate is quite fluctuating due to the reasons of inadequate irrigation facility and natural elements plays a vital role in overall production.

2.5 Review of Studies Related to the Constraints in Marketing

Gresta and lombardo (2008) studied that saffron being of great potential value and the considerable increase in new generation consumer demand for saffron, the future of the plant is still uncertain. Indeed, the main obstacles to saffron production are: (1) the limited areas of cultivation in countries where it is traditionally grown, (2) the great amount of sophisticated spice, (3) management techniques executed by hand, and (4) the very high price of the spice.

Nehvi *et al.* (2009) in his study found low productivity level of saffron in Kashmir (2.50 kg/ha) was a big challenge for the researchers and policy makers to ensure sustainability of this legendary crop. Lack of high yielding varieties, senile fields with inadequate plant population (2-3 lacs/ha instead of 5 lacs/ha) moisture stress (rain fed cultivation), inadequate availability of disease free saffron corms, nutrient depletion in saffron fields, longer planting cycle of saffron corms (>15years as against 4-5 years), higher incidence of pests and diseases, low saffron recovery up to 22g/kg of fresh flowers due to delayed flower picking and stigma separation, low colouring strength, owing to quality deterioration due to traditional drying practices (sun drying), inadequate quality control/certification/branding system, poor price discovery and lower farm gate involvement of intermediaries and adulteration and admixture were the major issues at present that resulted in decline in area (5707 ha to 3785 ha) and production as well from 16 M.T to 9.5 M.T in Kashmir

Nehvi *et al.* (2009) revealed that international demand and high prices were supporting factors to revitalize saffron cultivation. As per trade estimates, international demand is in the range of 300 MT per annum, while current production is in the range of 250-260 MT. hence, it was desirable to focus on productivity enhancement, improvement of post harvest processing and transparent marketing channels for an overall growth of saffron economy, and enhancement of income of saffron farmers.

Kumar *et al.* (2013) stated that Agriculture in India today is constrained by various factors. We have created history by producing 250 million MT of food grains in 2012-13, but this has been accompanied by land degradation, declining size of land holdings and many other related problems. While the strength lies in having the largest cultivable land with record food grains production, our weakness lies in having low yields, less value addition and food processing and large amount of post harvest losses. He also stated that where the opportunities existed and how the opportunities could be further strengthened to augment the yield and income of the farming community in the agriculture sector.

2.6 Review of Studies Related to the Support Service/Quantum of Finance

Rakesh (2004) reviewed performance of agricultural credit in India indicated that though the overall flow of institutional credit has increased over the years, there were several gaps in the system like inadequate provision of credit to small and marginal farmers, paucity of medium and long-term lending, etc. These have major implications for agricultural development as also the well being of the farming community. He, therefore, suggested that efforts were required to address and rectify these issues.

Giuliano *et al.* (2005) analysed that typical constraints faced by companies in developing countries include lack of specialized skills and difficult access to technology, inputs, market, information, credit and external services (finance) etc.

Campion (2006) suggested that value chain finance offers an opportunity to expand the financing opportunities for agriculture, improve efficiency and repayments in financing, and consolidate value chain linkages among participants in the chain. The specific opportunities that financing can create within a chain are driven by the context and business model and the relative roles of each participant in the chain.

Shwedel (2007) Despite the changes in agriculture and agribusiness, the typical offer for financial products and services for agricultural and rural production has been deficient and not particularly innovative; financial intermediaries still lack much depth in rural areas, and producers, especially smallholders, are still underserved. Conventional thinking is that the agricultural sector is too costly and risky for lending. Yet, major banks in the sector such as Rabobank and Banorte, large financial institutions in the Netherlands and Mexico respectively, both express the view that agricultural credit is profitable if producers are well integrated into a viable value chain.

Nyoro (2007) noted that in Africa 'value chain actors are driven more by desire to expand markets than by the profitability of the finance'. Traders, for example, commonly use finance as a procurement facility while input suppliers often employ it as part of a sales incentive strategy. For financial institutions, it offers an approach to lower risk and cost in providing financial services. For the recipients of value chain finance, such as smallholder farmers or those purchasing their products, value chain finance offers a mechanism to obtain financing that may otherwise not be available due to a lack of collateral or transaction costs of securing a loan, and it can be a way to guarantee a market for products.

Barah and Sirohi (2011) revealed that despite the significant strides achieved in terms of spread, network and outreach of rural financial institutions, the quantum of flow of financial resources to agriculture continues to be inadequate. As a result, agrarian distress on account of deceleration of two agricultural growths since late 1990's had been recognized as one of the major impediments in the development process of India. The adverse impact of such slowdown was more serious in the rain fed regions especially on small and marginal farmers with limited resources. Recent studies on agrarian distress have revealed that indebtedness is one of the factors linked with farmers' suicides on account of crop failure and related issues. This situation brings out the fact that the existing institutional arrangement for credit delivery is not adequate and suitable to address the agrarian distress in the country.

Puhazhendhi (2011) suggested that there was a significant need for increased institutional credit for agriculture; the Government of India initiated a series of policy measures since independence of the country. As a result the institutional credit structure in the country has shown a significant growth both in volume and complexity over the past few decades. At present there is an extensive banking infrastructure comprising 33,411 rural and semi urban branches of commercial banks, 14501 branches of Regional Rural Banks, around 12000 branches of District Central Cooperative Banks and nearly 1,00,000 cooperative credit societies at the village level which translates into at least one credit outlet for about 5000 rural people or 1000 households.

Ghabankandi *et al.* (2012) stated that exchanges have an important role in the economical development in most of the countries. The agricultural commodity exchange finances the producers of the agricultural sector. He stressed on providing effective responses to finance of saffron producers and to find solutions for their financial problems and information on the need of saffron producers to the agricultural commodity exchange as for improvement in the saffron markets. He suggested that agricultural commodity exchange needs a national attempt to achieve the goals and to have an effective role in the economy of this sector. Also method of financing and directing capital to agricultural sector was mainly discussed. Then by recognizing opportunities and threats of this market, it intended to provide the growth and development of Iran's capital market in agriculture sector in a scientific method.

2.7 Review of Studies Related to the Value Chain, Marketing Channels, Price spread and Marketing Efficiency

Garg and Azad (1960) found that the producer's share in consumer's rupee in marketing of oranges at Nagpur market during the year 1957-58 came to nearly 28 per cent, whereas middlemen like contractors and wholesale dealers earned as high as 19.3 per cent and 11.9 per cent respectively of the price paid by the consumer. The retailers earned a profit of Rs.91.75 per quintal, *i.e.*, nearly 10.8 per cent of consumer's price.

Wani *et al.* (1994) studied the economic viability of apple orchards in Kashmir. A sample of 160 apple growers was drawn randomly from selected villages. In order to find out the economic viability of apple orchards, the samples were collected in such a way that it contained orchards from all years of age. However, orchards of more than 43 years of age also were through exceptionally. The researchers found that the payback period is 14 years. The net present value was worked out to be Rs. 53417.14 ha-1, benefit cost ratio 2.29 and the internal rate of return 26 per cent. All these measures clearly revealed that the establishment of apple orchards in Kashmir is quite profitable and economically viable.

Toor and Poonia (1995) conducted a study on "Marketable surplus and price spread in the marketing of kinnow in Hoshiarpur district of the Punjab state." The study was taken up during 1990-91. It was found that the large farmers per cent contributed the major part of the product *i.e.* about 58 and the small and medium farmers together shared rest of the production. The average marketable surplus on small, medium and large farms worked out at 96.34 per cent, 97.72 per cent and 98.70 per cent, respectively of total kinnow production. The pre-harvest contract system was pre-dominant method of marketing kinnow fruits and about 70 per cent of the marketable surplus were sold through this system. The producer received maximum share of the consumer rupee (50.86 per cent) by selling the produce to Punjab agro-industries corporation.

Satihal and Hiremath (1995) in their study entitled "Price Spread and Marketing Margins for Ber in Karnataka : A Spatial Analysis" examined the marketing channels and ascertained the share of producer and the margins of intermediaries involved in the case of ber fruits in the three important markets, *viz.*; Bijapur, Hubli and Banglore of Karnataka state. Only one marketing channel was observed in all the three markets *viz.*, producer–commission agent cum wholesaler– retailer–consumers. The producer's share in the consumer's rupee was the highest in Bijapur market (58.82 per cent) followed by Hubli (49.33 per cent) and Banglore market (48.11 per cent) in spite of receipt of lower net prices of Rs. 487.92 q-1 by the producers in Bijapur market as compared to Rs. 505.65 and Rs. 553.27 q-1 received by the producer in Hubli and Banglore markets, respectively.

Singh *et al.* (1997) studied the "cost structure and economic potentials of horticultural crops in district Farrukhabad, Utter Pradesh". The study revealed that the cost structure of different enterprises showed a great variation on account of their input requirements. In the case of horticultural crops like guava the cost per hectare per annum worked out to Rs. 11,667, for mango Rs. 13,255 and for roses

Rs. 14,205 (average for the first three years). In all the three crops human labour accounted for the highest share (nearly 27 per cent) in their respective total cost. In the case of potato and vegetable crops group, the per hectare cost (C) was worked out to Rs. 25,920 and Rs. 8,110, respectively.

Pujari (1998) conducted a study on "Marketing of pomegranate and ber in Maharastra". He concluded the cultivation of pomegranate and ber covered full the working, capital, managerial and marketing costs and in addition, provided considerably higher than 20 per cent surplus over and above the aggregate cost of production. As such, growing of pomegranate and her has been quite a lucrative activity for the farmers in the drought prone areas.

Tripathi and Sharma (1999) conducted a study in Uttar Pradesh hills with objective to "Examine the Existing Marketing System of Apple". The system of marketing of apple included, picking, assembling, grading, packing and transportation. Picking was generally done manually. The average charges paid for picking, assembling, grading, packing and stitching were Rs, 13.47 for 18 kg wooden box and Rs. 13.29 for 60 kg gunny bagged fruits. The transportation was done only through road in the study area. The average freight from the forwarding point to the mandi was Rs. 5.23 and Rs, 14.17 per box and per bag.

Saraswat *et al.* (2006) studied the production of peach fruit in Rajgarh area of district Sirmour in Himachal Pradesh. The study revealed that the average maintenance cost of peach orchard was worked out to Rs. 65,227 in which per hectare fixed cost was Rs.43,299 and variable cost was Rs. 21,928. The higher fixed cost was due to higher prorated establishment cost. Peach production was found to be economically viable on all size of farms. Overall per hectare net return were worked out to Rs. 8558, which were Rs. 2347, Rs 6117 and Rs 12734 on marginal, small and medium size of farms, respectively.

Anchal and Sharma (2009) conducted a study was "Price spread of litchi in Punjab". The present study was undertaken in Gurdaspur district of Punjab with a view to examine various marketing channels and price spreads of litchi in the study area. The primary data from 40 litchi growing farmers have been collected from two randomly selected blocks. The results of the study showed that the producer revived the maximum share in the consumer's rupee whereas the consumer paid the minimum price when the litchi was sold as well bought in the local market. The margins of the middlemen between producer and consumer was found to be at the tune of Rs. 255.80 in the channel producer –retailer – consumer in local market, Rs. 524.80 in the channel of producer – pre-harvest contractor –retailer – consumer in Amritsar market and Rs. 910.49 per quintal in the channel producer – pre-harvest contractor- retailer – consumer in Delhi market. This increase in the margins of middlemen and costs of litchi per quintal led to the higher price spread in the marketing of litchi form local to Amritsar and further to Delhi market. Thus from the economic point of view the local market was the most suitable channel in respect to litchi growers of Punjab.

Johl and Thakur (1968) in their paper entitled, "Marketing of Apples in Himachal Pradesh" focused on marketing margins and costs in the chain of distribution of apples in H.P. The study revealed that sale through commission agents (45 per cent of the produce sold) was the most commonly used channel. Sale to wholesalers, sale to contractors, sale direct to consumer, sale through co-operatives and sale to retailers accounted for 32.8 per cent, 14.3 per cent, 6 per cent, 1.5 per cent and 0.4 per cent respectively of the total produce marketed. Net earnings in case of direct sale to consumer were about twice the contractual system and 1.3 times more than sale in the market through other existing methods of apple marketing. The study concluded that the distribution costs due to handling and preparing the produce for the market assembling, storage and transportation was 29.79 per cent of the consumer's rupee.

Raghubanshi and Dhall (1971), in their study, have examined price spread for green hill peas in Solan district of Himachal Pradesh in different marketing channels in the year 1968-69. The study reported that the producer's share was as low as 60 to 65 per cent of consumer's rupee in the channels where wholesaler, retailer and the commission agent were involved. For narrowing the price spread and for increasing the producer's share in consumer's rupee, the study suggested the elimination of wholesalers and also reduction in commission agent's commission from a level of 6 per cent to 2.5 per cent.

Singh (1974) identified the agencies involved in the trade of grapes, cost of marketing at different stages of marketing of grapes and share of producer in consumer's rupee in Hyderabad. The study concluded that the marketing charges were found to be less by Rs.4.35 for hundred rupee worth of sale of grapes in a regulated market than in any unregulated market. Marketing cost of retailer was worked out to be 1.97 per cent of consumer's rupee in both the channels. The commission charges were worked out to be 3.96 per cent of consumer's rupee in both the channels. The producer had a share of 47.18 per cent and 69.63 per cent in the price paid by the consumer in the channel-I and channel-II respectively. Obviously channel-II was more profitable than the channel-I to the producer.

Patel and Pawar (1980) studied the marketing of fruits in Mahatma Phule market, Mumbai. The study showed that there was close relationship between wholesale price and supply position, whereas there was no such pattern in case of retail price. The producers share in consumer's rupee was 33.07 per cent for sweet oranges. In case of apples, mangoes and grapes, the share of the producer in consumer's rupee varied according to the varieties and ranged from 33 to 54 per cent. The study also made some suggestions as standardization and grading, and establishment of cooperative marketing societies for getting better prices for the producers.

Authurkar and Deole (1985) estimated the producers share in the consumer's rupee for banana, sweet orange, mandarin orange and sour lime in the Marathwada region of Maharashtra and reported that it ranged between 28 per cent and 30 per cent. The margin of the intermediaries accounted for about 24 per cent of consumer's rupee while about 46 per cent to 48 per cent of the consumer's price went towards marketing costs.

Mondal (1986) conducted a study on "Marketing of pine apples in Meghalaya state" and analysis of price spread showed that net profit of the producer was highest (88.32 per cent) when the produce was sold directly to consumers. Net percentage margin received by the retailer when operating through 'producer-retailer-consumer' channel was 28.70 to 30.19 per cent, whereas, net share of wholesaler was 14.84 per cent to 15.20 per cent.

Shah (1986) stated that Marketing problems of apple industry in Kashmir valley. Jammu and Kashmir is the largest apple producing state of India, but the state does not earn that from the apple industry what it should have. The main problem of apple industry in Kashmir is the lack of marketing information and techniques. Without a proper marketing information system the market opportunity cannot be fully exploited. Growers should be provided proper market information to get dispose off their produce in those markets where it can yield maximum prices.

Porter (1990) concluded that factor conditions relate to the nation's endowment with resources such as physical, human, knowledge, technology and infrastructure. These factors enable or constrain value chain upgrading.

Christopher (1992) suggested that the value chain was the core business process in an organization that created and delivered a product or service from concept through development and manufacturing into a market for consumption and the integrated supply chain consisted of suppliers and organization.

Bannerjee *et al.* (1998) revealed that in the long-run, even a one per cent increase in physical capital in Saffron leads to a 0.45 per cent increase in agricultural value added. While, a one per cent increase in human capital in Saffron leads to a 0.22 per cent rise in agricultural value added. Similarly, a one per cent increase in oil exports leads to a 0.14 per cent decrease agricultural value added. Moreover, empirical results show that a one per cent increase in Saffron exports leads to 0.35 per cent increase in agricultural value added. It is obvious that Saffron exports have an effect on the Iranian economy which, though statistically significant, is more so than expected.

Vitonde *et al.* (1991) studied the marketing of mandarin oranges in Nagpur district of Maharashtra. They worked out the indices on the basis of prices during the period of 10 years. *I.e.*, from 1976-77 to 1985-86. They had seen that the prices were high in the month of December, January and April as compared to other months and that was due to reduction in supply of oranges in relation to demand. The lower prices existed in October and May due to low quality of produce.

Christopher (1992) suggested that the value chain was the core business process in an organization that created and delivered a product or service from concept through development and manufacturing into a market for consumption and the integrated supply chain consisted of suppliers and organization.

Singh and Sikka (1992) revealed that during marketing in tribal areas of Himachal Pradesh, producers share in consumer's rupee was 33.29 per cent. The marketing cost borne by orchardists was 41.07 per cent and margins of mashakhars and retailers were 2.33 per cent and 11.85 per cent of consumer's rupee, respectively.

Iqbal (1992) observed that the gross margins of producer, pre harvest contractor, wholesaler and retailer were 13.11 per cent, 30.90 per cent, 12.51 per cent and 43.48 per cent for plums in Baluchistan, respectively while the margins of pre harvest contractor, pharia and retailer for apple were 32.65 per cent, 9.90 per cent and 26.99 per cent, respectively. The producer share was 38.65 per cent and the net commission agent share was 3.95 per cent. The net margins of almond for producer, pre harvest contractor, wholesaler and retailer were 25.2 per cent, 24, 10 per cent.2 per cent and 37.2 per cent, respectively. He noted that the producer of almond who self marketed were able to improve his share from 25.2 to 45 per cent while the commission agent charged the commission at the rate of 2 per cent of the auction price.

Singh and sikka (1992) revealed that during marketing of apples in tribal areas of Himachal Pradesh, producers share in consumer's rupee was 33.29 per cent. The marketing cost borne by the orchardists was 41.07 per cent and margins of mashakhars and retailers were 2.23 per cent and 11.85 per cent of consumer's rupee, respectively.

Tomer *et al.* (1997) studied the marketing cost and margins for citrus (malta and kinnow) in Hisar and Sirsa districts of Haryana. The study revealed that producers share in consumers rupee was around 50 per cent when the producer directly sold citrus in the market, however, when sold through pre harvest contractors, the share declined to about 40 per cent. The marketing margins charged by the middleman for citrus were higher which ranged from 14 to 18 per cent of the consumer price.

Shiyani (1998) highlighted that the marketing of vegetables has more problems as compared to other agricultural commodities as they have high degree of perishability, bulkiness, higher proportion of retailer's margin and concentration of trade in few hands. The analysis revealed that in South Saurashtra zone of Gujarat, the overall marketed surplus was more than 90 per cent of the total vegetable production. The commission charges, transportation cost, spoilage cost turned out to be the most important components among all the items of marketing costs. The producer's share in the consumer's rupee ranged from 56.87 per cent in tomato to 62.38 per cent in cabbage.

Khair (2000) observed that overall marketing margins for apple varieties were 73 per cent while the producer was receiving 27 per cent of the consumer price. The marketing cost was very high due to exploitative marketing set up, lack of marketing intelligence of farmers, expensive packaging material and transport.

Kaplinsky (2000) suggested that the value added is created at different stages and by different actors throughout the value chain. Value added may be related to quality, costs, delivery times, delivery flexibility, competiveness, etc. The size of value added is decided by the end-customer's willingness to pay. Opportunities for a company to add value depend on a number of factors, such as market characteristics (size and diversity of markets) and technological capabilities of the actors. Moreover, market information on product and process requirements is key to being able to produce the right value for the right market. In this respect finding value adding opportunities is not only related to the relaxation of market access constraints in existing markets but also to finding opportunities in new markets and in setting up new market channels to address these markets.

Lazarrini *et al.* (2001) stated that the network structure has two dimensions: vertical and horizontal. The vertical dimension reflects the flow of products and services from primary producer up to end-consumer (*i.e.* the value chain or supply chain). The horizontal dimension reflects relationships between actors in the same chain link (between farmers, between processors, etc.) developed the concept of the net chain to show the interrelationships between the horizontal and vertical dimensions in value chains Channel choices are heavily constrained by market access limitations such as supporting infrastructures to reach markets, access to demand and price information and specific demands from these markets such as production according to quality standards. Moreover, the ability of companies to take part in market channels is strongly related to characteristics of these markets, knowledge of market demands at the producer and the technological abilities of the producer.

Zaki *et al.* (2002) stated that Agricultural marketing in its wide sense comprises all the operations involved in the movement of produce and raw material from the farm to the final consumer. It includes the handling of the product at the farm, initial processing, grading and packaging in order to maintain and enhance the quality and avoid the wastages. Marketing of the saffron is concentrated in the hands of a few traders because a common grower cannot directly sell his meager produce as he cannot grade, pack and store the produce at individual level. Because of the poor financial status the farmers are prompted to sell their produce through middlemen to the owners of the trading firms. Between the grower and the ultimate consumer there exists a long chain of the intermediaries.

Sharan and Singh (2002) examined the marketing problems faced by kinnow growers in Rajasthan. The study revealed that selling of produce through self marketing by growers was found profitable in comparison to contract sale to pre harvest contractors. the major problems highlighted by the study were lack of support price, lack of growers organization, delay in payment, lack of marketing information, lack of cold storage facilities and lack of better and cheaper packing material. The growers also reported other problems like lower price due to seasonal gluts, lack of stay arrangements in the market, malpractice in weighing etc.

Ladaniya *et al.* (2003) examined the price spread of pomegranate. The study highlighted that pomegranate was mainly grown in Maharashtra, Karnataka ad Rajasthan in the country. Over 90 per cent of produce was marked in three production centers of Sangola, Malagaon, Rahuri through the following major channels: (1) producer commission agents → wholesale → retailer → consumer, (2) producer → cooperative society → commission agents → retailer → consumer (3) producer → commission agents (local) → trader → wholesaler → retailer → consumer. Local channels *viz.* (i) producer → consumer (ii) producer → retailer → consumer (iii) producer pre-harvest contractor → retailer → consumer were also followed through which 10 per cent of the produce was marketed. Producers with small holdings preferred cooperative society and marketing of produce through forming self help groups for transportation. Packing long distance transportation and commission charges accounted for 90 per cent of the marketing costs. Retailer's margin was in the range of 38.50 to 56.33 per cent in the price paid by the consumer.

Ladaniya *et al.* (2003) studied the marketing pattern of mosambi in selected districts of Maharashtra and observed that mosambi marketed through three different channels such as (1) Farmers → pre harvest contractors → commission agents → retailer → consumer. (2) Farmers → commission agents → wholesalers → retailers → consumers. (3) Farmers → pre harvest contractors → commission agents → buyers of distant market → wholesaler → retailers → consumers. Commission and transportation charges mainly contributed towards cost of marketing incurred by producer. The maximum increase in the retail price of the mosambi was due to retailer's high margin.

Kasana (2003) carried out study on distributive marketing margins of three most commonly grown vegetables, *i.e.* potato, peas and marrow and the shares of different marketing functionaries involved in the marketing margins. He observed that total marketing margins for potato was 38.86 per cent, for peas 54.89 per cent and for marrow 62.89 per cent. The net margins for potato, peas and marrow were 19.04 per cent, 27.25 per cent and 30.50 per cent respectively. The producer received 61.136 per cent, 45.106 per cent and 37.107 per cent of the price paid by the consumer for potato, peas, and marrow respectively. The difference in marketing margins for various vegetables was due to high marketing and picking costs. It was observed that 30 per cent of the potato fields were sold to pre harvest contractors. The highest marketing margins were observed for marrow followed by peas and potato respectively. The highest net margins for producers were observed for potato followed by peas and marrow. The highest net margin for wholesales was found in marrow followed by peas and potato. The retailer's highest net margins were observed for marrow followed by peas and potato.

Hosseini *et al.* (2003) examined that Saffron is a strategic product of Iran. Iran's share of Saffron from the world production and export was almost 90 per cent. Labor requirement for Saffron production was 200 man day per hectare. However, this figure was much higher in the processing and marketing sectors. DRC and SCB indices showed that production of Saffron in Iran has comparative advantage. Marketing margin of Saffron was high so that producers received less than 65 per cent of final price of consumer. According to special advantage of this product, creation of regional marketing board besides agricultural exchange market for coordinating production, marketing and export and maintaining market share was suggested.

Chopra and Meindl (2004) elaborated a supply/value chain design problem comprising the decisions regarding the number and location of production facilities, the amount of capacity at each facility, the assignment of each market region to one or more locations and supplier selection for sub- assemblies, components and materials so as to enhance value maximization at every end.

Ajani (2005) estimated the marketing margins, net margins and profitability at different levels of marketing. Results revealed that tomatoes had highest marketing and retail level. This implied a wide gap in prices between wholesalers and retailers, the study also revealed that fruits were cheapest in oyingbo market. The least marketing margin was recorded in the sales of bananas at the wholesale level (19 per cent) and oranges (18.09 per cent) at the retail level. There was a higher marketing efficiency in some fruits (banana) than another (tomatoes) because a higher percentage marketing margin showed lower marketing efficiency.

Grunert *et al.* (2005) analysed that the more heterogeneous and dynamic the supply of raw material to the value chain, the more market-oriented activities can be expected to take place upstream in the value chain. Conversely, from an end-user market perspective, they find that the extent of heterogeneity and dynamism of end-user markets is a determinant of the degree of market orientation in the chain.

Muiruri (2007) stated that the value chain concept allows integration of the various players in agriculture production, processing and marketing. It defines the various roles of players while at the same time, scope and purpose of partnerships that can be established.

Mayoux and M (2007) stated that value addition is a holistic approach because it pays attention to the complex interactions of income, value added across the chain and how these are distributed within particular points of the chain and across the different levels of the chain.

Kaur and Singh (2007), in their paper entitled, "Price Spreads and Marketing Efficiency of Kinnow in Sri Ganganagar District of Rajasthan.A Temporal Study" have examined the prevailing marketing channels, margins and price spreads of kinnow in the study area for the year 2002- 03 and suggested some policy measures for the improvement of existing system. Two types of channels of kinnow marketing were studied: (i) Producer → Pre-harvest contractor → Consumer (ii) Producer → Wholesaler → Retailer → Consumer. The results of study revealed that the practice of pre-harvest contract was adopted by considerably a large number of orchardists. From consumer's point of view, channel-II was the better distant channel as the consumer got kinnow at the lowest price in this channel. However, the practice of pre-harvest contracting definitely reduced the actual profits of the orchardists and discouraged them from producing the crop on large scale. The study suggested that efficiency of marketing system should be improved so that the producer has a better stake in the consumer's rupee involving this commodity.

Nehvi and Wani (2008) examined the price spread of saffron in various marketing channels. The channel-1 (producer-consumer) was followed by 4.00 per cent of the selected respondents. The producer after incurring marketing costs of Rs 73 per 10 gram and Rs 10 per 10 gram got a share of 85.40 per cent and 97.50 per cent in consumer's rupee for grade-1 and grade -11 of saffron respectively. The major item of cost in grade-1 of saffron was the expenses on grading.

Sarhad (2008) studied distributive marketing margins of rice and the shares of different marketing functionaries involved in the marketing margins in Batkhela Tehsil of Malakand district during the year 2004. It was observed that two marketing channels 1) Producer, wholesalers (Pharia) retailer, consumer and 2) beopari, wholesaler (Pharia) retailer, consumer, involved in trading of rice in the study area. In channel 1, the producer received 17.90 per cent net margin and 41.04 per cent gross margin. However, in channel 2, it was found that the producer gained net margin of 36.36 per cent and 14.54 per cent gross margin. Furthermore it was also observed that the lack of capital, poor extension services, high input price and lack of marketing channels were the main marketing problem of rice producers in the study area. Additionally total production, marketing intelligence, education, marketable surplus and marketing price were important variables affecting marketing margin.

Gibbon *et al.* (2008) in the contribution concluded that global value chains are characterized by falling barriers on international trade due to decreasing tariffs and the lowering of price support and export subsidies in the last decades. At the same time we see increasing concentration and consolidation in all links of these chains. Furthermore, advances in communication technologies and declining transportation costs facilitate coordination between chain actors.

Shah *et al.* (2009) the study revealed that the economics of saffron production, the gross expenditure of crop during 5 years planting cycle of saffron was Rs 43,1,185 per hectare of land because of high establishment cost in first year and the gross returns in the planting cycle of 5 years were Rs 2,36,6,000. The net profit per annum was calculated Rs 3,86,963/ha. if saffron was sold @ Rs 230/gm and average net profit per year per kanal of land was Rs 19,348.15 including seed corms which were obtained after completion of 5 years were projected for Kashmiri saffron.

Rubin *et al.* (2009) described value chain analysis (VCA) as 'the process of documenting and analysing the operation of a value chain, and usually involves mapping the chain actors and calculating the value added along its different links

Gammage (2009) stated that value chain analyses provide opportunities for showing that various value chain actors may influence capabilities of other actors, possess different levels of bargaining power and subsequently affect outcomes along the value chain.

Herr and Muzira (2009) observed that monitoring and evaluation are essential processes in value chain development. It is through them that implementers and facilitators can gauge if their interventions are on track and are achieving the desired goals. A monitoring system is important because it tracks activities against set targets periodically and shows where there is need for corrective action. It looks at the financial, human and material resources used (inputs), the products, goods and services which result from development intervention (outputs), the likely or achieved short- and medium-term effects of intervention outputs (outcomes) and the achievements of overall long-term and strategic objectives which could be the expected or unexpected positive or negative (impacts) of development initiatives.

Laven *et. al.* (2009) also outline an integrative framework for value chain and gender analysis by integrating two separate frameworks on gender empowerment and chain empowerment to provide insights into the internal dimensions of value chains such as vertical and horizontal integration.

Riisgaard *et al.* (2010) suggested that Value chain analysis is also perceived as a means of understanding trade at the global level as well as strengthening systemic competitiveness. It identifies vertical and horizontal components in a system of stages/nodes of physical transformation processes that are inter-linked by transactions that occur either in the same firm or between firms in similar or different geographic locations

Riisgaard *et al.* (2010) noted that adopting the value chain approach as a development strategy provides an opportunity for all actors to understand each other's functions and the activities involved; increase their viability, visibility, voice and market share; and identify and correct barriers and gaps that cause inefficiencies. Corrective value chain interventions aim at creating or enhancing horizontal relationships (among actors within the same level in value chains) and/or vertical relationships (among actors in different levels of a value chain) with an aim of improving returns and increasing efficiency. They may include formation of new value chains, forging or strengthening new links within an existing value chain, increasing the capabilities of target groups to improve the terms of value chain participation and minimizing the possible negative impacts of value chain operations on non-participants and/or adjacent communities.

Bagheri *et al.* (2011) stated that the main purpose of the study was to provide awareness of different rings of chain supply model and analysis entry kind, reasons of entry or don't entry of Iranian producers and exporters firms of saffron to this chain. this study is intended by using trade off cost economy so that in first step, survey on theoretical basis and other studies about this field, also was designed questionnaire for one of model of Porter (the five competitiveness factors; demand condition, factors condition, services condition, firms structure in competition and government role. The aggregated information led to design of chain supply model for saffron of Iran. The results showed that with the existence of total limitations trade off cost economy could be used for analysis of organizations approaches of chain supply model and the end led to decreasing export share and for competitiveness power for saffron of Iran in export of world markets.

Lawal *et al.* (2011) examined value addition to cashew as a way of preventing farm losses due to wastage and lack of proper storage of the cashew apple in Nigeria. The process of value addition involved the kernels being graded, heat treated, shelled roasted and packaged. The apples were crushed, processed to juice and bottled for sale. There was a significant difference ($P<0.05$) between net income per farmer adding value (US$487.26). Also the benefit-cost ratio of adding value was 1:2.30.

Mahmoud *et al.* (2012) stated that the structural equation modeling approach is widely used to analyze relationships and causation among manifestly observed and intrinsically latent variables, and control observation or measurement errors in economics, sociology, and psychology. The main objective of this study was to investigate the effects of three variables e-commerce, brand and packaging on the value added of saffron by using a structural equation model. Results showed that packaging parameter with regression weight of 0.91 had the greater correlation to the latent variable of "factor" and it showed that it had the share of high value added to two other scales. Also based on this result, it could be said that three scales together had more value added in the foreign market (0.35 coefficients) towards domestic market (0.26 coefficients).

Jacques *et al.* (2012) stated that for developing country value chain analysis comprised of three components. The first consisted of identifying major constraints for value chain upgrading: market access restrictions, weak infrastructures, lacking resources and institutional voids. In the second component three elements of a value chain were defined: value addition, horizontal and vertical chain-network structure and value chain governance mechanisms. Finally, upgrading options were defined in the area of value addition, including the search for markets, the value chain- network structure and the governance form of the chain.

Mehdi *et al.* (2012) stated that in recent years, appropriate increase of production for saffron export has led to seeking new export markets in Iran. The purpose of this study is to determine the relationship between Saffron export and agricultural value added in Iran. The theoretical framework was designed based on this assumption that the total production in the economy is divided into two sections: production for inside and production for export. The data were collected from 1990 to 2007 and were analyzed using Auto Regressive Distributed Lag (ARDL) model. The result of the analyses showed that there was significant relationship between Saffron export and agricultural value added. Together the independent variables explained 91 per cent of the variance in the dependent variables. The remaining 9 per cent was due to unidentified variables. In relation to that, we can conclude that explanatory power is high for the equation. It showed that one percent change in Saffron export rate led to 35 per cent in agricultural value added growth. Therefore Saffron export is regarded as an important factor in Iran's agricultural value added.

Kheirandish *et al.*, (2012) examined the marketing efficiency and price spread for saffron in Iran and also to identified the major problems thereof. It was based on the primary data collected from 50 traders (10 local traders, 12 wholesalers, 10 retailers, 11 processors and 7 exporters) in the markets of Mashhad, Torbat, Heydarieh and Ghaen were selected randomly. The secondary data were collected from published sources in Iran such as, Ministry of Jihad-Agricultural, Trade Promotion Organisation of Iran, Customs, ITC, European Commission, etc. Apart from using the usual techniques/concepts for analysing the marketing costs, marketing channels, marketing margin and price spread, Shepherd's Index was used for computing marketing efficiency. The study revealed that the maximum marketing margins were taken by the middlemen followed by wholesalers in Mashhad for bulk export to Spain, UAE and Italy traders. The packaging and processing companies, which usually performed the task of grading, incurred the maximum marketing costs. The gross price spread was highest (96.25) per cent in the direct marketing channel namely: Producer →domestic consumer and least (17.76) per cent in the channel: Producer → middlemen → foreign traders → foreign consumer. Altogether, 11 channels were identified in the marketing of Iranian saffron and the average share of producer in the consumer's price in the all the 11 channels was 51.23 per cent. The study revealed that there was considerable scope to increase the producer's share in the consumer's price if the number of intermediaries were reduced and the government intervened pro-actively in order to organize and streamline the marketing cooperatives unions so that the farmers use these unions as a profitable channel to sell their product. There is no effective control on intermediaries in the saffron market, at percent. The problem of irregular supply can be solved by forward contracts to be signed between producers and marketing bodies. The prices of saffron witness frequent fluctuations and even day to day fluctuations causing concern to the farmers.

Bhat *et al.* (2012) studied about the marketing efficiency of Kashmir apple. Marketing efficiency is important for increase in production and fair returns to apple growers. Marketing efficiency is measured in terms of price spread. Lesser price spread means more marketing efficiency and vice versa. They talked about three marketing channels and have concluded that marketing channel *i.e.,* Grower to consumer is having less price spread and more returns to growers but is rare in practice due to lack of marketing information, credit and institutional facilities, small holdings.

Prabhavathi *et al.* (2013) stated that India 'Land of spices' is the major producer and exporter of chillies. An efficient supply chain ensuring remunerative prices to the producers for their products and to deliver maximum satisfaction to the end consumers for the price they pay. Two major supply chains have been identified, which revealed that supply chain II was more efficient than the supply chain I because more value goods were delivered to consumer from producer at low marketing costs. The study showed that farmers who brought good quality chilies to market preferred Supply chain-II to supply chain-I. But farmers who brought poor quality, discolored chilies, were preferring supply chain-I.

Chapter 3

Value Chain Analysis of Saffron in Kashmir Valley of J&K State

3.1 Introduction

Saffron falls under the category of spices and the spice board of India (The Spices Board of India (Ministry of Commerce, Government of India) is the apex body for its export promotion. Established in 1987, the Board is the catalyst of these dramatic transitions. The Board plays a far reaching and influential role as a developmental, regulatory and promotional agency for Indian Spices. Spices are strongly flavoured or aromatic substance of vegetable origin, obtained from tropical plants, commonly used as a condiment". Spices were once as precious as gold. India plays a very important role in the spice market of the world. India produces a wide range of spices. Having varying climates from tropical to sub-tropical to temperate almost all spices grow splendidly in India. Almost all the states and union territories of India grow one or the other spices. Under the act of Parliament, a total of 52 spices are brought under the purview of Spices Board. However 109 spices are notified in the ISO list. The Indian spices can be categorized into three main categories as basic, complimentary and aromatic or secondary spices (Saffron). At present, production is around 3.2 million tonnes of different spices valued at approximately 4 billion US $, and holds a prominent position in world spice production. With global demand for spices and spices products rising, India is targeting exports of 2.3 billion US dollars in 2013-14 fiscal even as Rs 9,433 crore worth spices were shipped during April-December last year. The total volume of spices and spices products exported in the nine month period of April-December 2013 was 5,71,680 tonnes, valued at Rs 9,433 crore, a 41 per cent growth in rupee terms and 27 per cent both in volume and dollar terms. During the same period of the previous fiscal, as much as 4, 49,926 tonnes valued at Rs 6,696 crore ($1232 million) was exported. For 2014-15 fiscal also the board was targeting spices exports of $2.3 billion. (Spice board of India). Among the spices, Saffron (crocus sativus) is a perennial herb which belongs to the Iridaceae family and is the most expensive species in the world for its aroma and colour. It is the world's most favored species and its monopolistic character makes it more and more expensive. The three stigmas of the saffron flower are the most important economic part of the plant. The saffron stigma is rich in aroma and colour. In dried or powdered forms, stigmas are commonly used as "Spice or colouring in food preparation, Materials in pharmaceutical, cosmetic and perfume industries and Dye material in textile production.

The countries like Iran, Spain and India are the major saffron producing nations of the world with Iran contributing to the 88 per cent of the world's saffron production and at the same time India contributes around 7 per cent of the total production with the average productivity of 2.30kg/ha (Anonymous). Production of saffron in India is restricted to the states of Jammu and Kashmir and Himachal Pradesh with the area of around 4265 hectares and annual production 7.50 MT. out of this about 2469.02 hectares lie exclusively in Jammu and Kashmir State.

In Jammu and Kashmir which is considered as the second largest contributor of saffron to the global market, the cultivation is confined to district Pulwama, Budgam and Doda (kishtwar). District Pulwama accounts for 75 per cent of total area under saffron in the state followed by the District Budgam accounting for 16.13 per cent of the total area. District Srinagar accounts for 6.68 per cent whereas, Poochal, Namil, Cherrad, Hullar, Blasia, Gatha, Bandakoota and Sangrambatta of District kishtwar account for 2.5 per cent of the total area of the state. As per the official sources in the past 10 years there has been a steady decline in saffron production due to shift of agricultural land to the commercial purposes, in 1998 the crop was grown over 4161 hectares which has come down to 2469.02 hectares in 2011-12 which is the threat to the saffron industry in J&K. However state and central government and some research intuitions have taken some important steps to boost the value of Kashmiri saffron with the world bank launching a project called "value chain on Kashmiri saffron" which aim to boost production and quality using environment friendly techniques, and there is being demonstration of plots in J&K and national saffron mission to promote the saffron production in Jammu and Kashmir. Apart from this government has promised of providing 70 Kanals of land for setting up the saffron research centre to fetch a good amount on the account of exports. Low prices and competition from cheaper imported Iranian saffron have forced many traditional saffron farmers to abandon this crop so government need to step in urgently to protect this sector before it is too late. With the passage of time the saffron of Kashmir is dying that can be guarded/noticed from the fact that its production has decreased from 15.59MT to 10MT.

3.2 Value Chain of Saffron

Saffron is an ideal value chain for analyzing agricultural finance. Saffron is a profitable product for all value chain participants. As saffron production matures over a five or six year period, and due to relatively high establishment costs, saffron requires significant working and investment capital. Saffron market transactions are based on established relationships. There are two main saffron channels viz: Agents/ Commission Agents, Wholesalers and Retailers. They directly buy saffron from the saffron famers/producers in villages where they have long standing connections. The primary constraint to expanded saffron production is corm (planting material), supply good quality, adequate quantity, and right price. Selling saffron corms is a major product for saffron farmers. Prices for corms increase every year due to inadequate supply.

Value chain actors in Kashmir comprises of saffron farmers agents/input suppliers, wholesalers and retailers and ultimate consumer. Each and every actor has its vital role in the saffron business. There exists a type of interdependency between the chain actors so that is why these actors are the indispensible determinants and elements of the saffron business as well as in the saffron cultivation in Kashmir. Every actor has a role assigned as per their need and requirement. The chain comprises of:

3.2.1 Input Supply

There are various Actors involved in support services for saffron cultivation in Pulwama district. The Government, SKUAST and KVKs are the supporting organizations which have been working on extension activities for Saffron extension in Pampore in District Pulwama. In the inputs supply level of the subsector, provision of the saffron corm is the biggest concern of the saffron producers in Pampore. Among all involved parties in input supply for saffron, the farmer is the biggest supplier of corm. Government has established 2 corm farms and provides corm to the farmers at subsidized rates. Govt has also provided saffron dryers to the Saffron Producer' Association and has conducted training workshops on saffron cultivation, harvesting and post harvesting of saffron

3.2.2 Producers

Although saffron cultivation and production is an age old activity in Pulwama district, but somehow it has not been scaling up. A good yield from a kanal of saffron land in Pampore is near about 1 kg/kanal, To make one kg of saffron it takes some 450,000 stigmas and as a saffron flower has only 3 stigmas therefore the workers need to process more or less 150,000 blossoms to produce a kg of saffron stigma. Thus it is also labour generating.

3.2.3 Processing

Collection of saffron flowers takes place in the early morning and they are transported to the farmhouse or to a process centre, where the flowers should be ideally kept in a clean and shady place until the stigmas are separated from the blossoms. The collection of flowers and separation of stigmas are performed by hand. About 80 per cent of flower collection and stigma separation is performed by women who are working on a daily wage and get paid Rs 350 per day. If the processing of the flower is delayed it is necessary to store the flowers at temperature of about

0°C, because wilting of flowers makes the processing difficult or even impossible. On the other hand it will also decrease the quality of stigma. As soon as the stigmas are separated from the blossoms it should be dried immediately and in a proper way to keep the quality and retain the required moisture. If the stigmas are too dry it will break easily and turn into powder. On the other hand if it is too moist it may spoil or get infected with fungus. Although government have provided dryers to the farmers but they are not enough and therefore the farmers still use traditional sun drying method which takes a week to dry the stigmas.

3.2.4 Packaging

After the saffron is processed it should be packaged and sealed in proper containers. It is also required to keep the saffron away from the direct sun light. Tins and dark glasses are good to store the saffron for a longer time. It is also possible to keep the saffron in clear glasses containers to be visible for quality assessment by customers but it should be stored in a dark place until it is sold. Some countries like USA would like to receive the saffron in bulk and packaging it in their own standards, (Malik, and Wyeth, 2008). But some other countries like Dubai prefer to receive the saffron packaged. Currently there are no such big players or companies indulging in the practice of packaging in district Pulwama. No company is registered with the District Industries Centre Pulwama for packaging or processing of saffron.

3.3 Importance of the Study

Kashmir has the potential of becoming a global leader in the saffron industry through the adoption of a scientific agro-technology, better post-harvest management and proper marketing strategy. Apart from factors related to agronomic production practices and lack of scientific aptitude among small and marginal farmers (Husaini *et al.*, 2010), clandestine import of Iranian saffron and rampant adulteration practices have ruined the saffron market of Kashmir. Inadequate infrastructural facilities and poor quality control measures have shaken the confidence of growth and customers besides long value chain of marketing. Replacement of marketing systems by a new intervention have made it possible for Iran to export about 120 tonne of saffron compared to 1.3 tonne exported from India (2005-06) (Nehvi *et al.*, 2012) where farmers are dependent on middlemen, traditional sun drying and family labour. It is well established that the demand of quality saffron in the world market is enormous and therefore linking the growers with the world market traders is an essential task. Although the government organizations from time to time have played important role for boosting its production and market linkages but still there is ample scope to study and channelize the market keeping in view of earning a good international price which otherwise may play a leading role in earning a good foreign exchange for India. Evaluating the economics of marketing will help the saffron growers or farmers of this region to a greater extent as how to make their marketing more profitable besides will also act as a guideline for the planning of policy planners/researchers.

Saffron Packing.

3.3.1 Objectives of the Study

1. Understanding the status and importance of the selected crop (saffron)
2. Identify the value chain model that currently exists for the saffron crop.
3. Identify the interest and economic relationship of stakeholders- the level of dependence and relationship between them.
4. Assess the value added in the various levels and the transactional flows of the product within the chain.
5. Assessing the supports service including quantum of finance, available and needed at the various points of the chain- technical assistance and government support at various stages of the value chain, regulatory constraints and potential support from the government or the organizations, the key threats and opportunities in the entire value chain.
6. Identifying the key determinants of the share of the profit created by the chain etc.

3.4 Scope and Limitations of the Study

3.4.1 Scope

- Study of marketing cost components and price spread in different marketing channels will help to identify the producer's share in consumer's rupee and also most efficient channel.
- The study has indicated the possible measures taken at various levels of the chain activities in order to make the crop more viable and value added.
- The study also has indicated the public as well as the private role in quantum of finance provided to the chain actors for the production, marketing and value addition purpose.
- The study has indicated the possible measures to make the marketing practices most efficient.

3.4.2 Limitations

- The study was based on sample survey of selected saffron farmers from major saffron growing areas of Kashmir region. The information was collected by personal interview method based on respondent's remembrance, past experience and his involvement. Though maximum possible efforts were made to collect reliable information, yet there may be some lacunae in the information given by the respondents.
- The field study has been confined to Pampore tehsil of district Pulwama. It cannot pretend to generalize the results for the Kashmir valley but it may be helpful as a guideline for further studies.

☆ Owing to time and resource constraints, a limited size of sample (95) was taken. A larger sample size would definitely tend to improve the reliability of the result.

3.5 Methodology

The sampling structure and techniques adopted during the course of investigation have been described in this chapter.

3.5.1 Locale of Study

The Jammu and Kashmir state is divided into three regions-Jammu, Kashmir and ladakh. The Kashmir region of the state was selected purposively as per the requisition of NABARD whose project was undertaken by the researcher. Thus Pulwama district was selected purposively on the basis of area of saffron under cultivation *i.e.*, 75 per cent out of the total area under this crop

3.5.2 Collection of Data

The primary data from growers of saffron was collected by survey method, using pre- tested structured schedule. Collection of the data was done by the personal interview method of saffron growers/farmers, various markets and contacting the different intermediaries involved in marketing and disposal of the saffron spice. Also to study the various support services or market related information, various market functionaries, commission agents, wholesalers and retailers were selected randomly and the required information as well as data was obtained from them. Secondary data was collected from various published sources such as bulletins of the Ministry of Agriculture, Govt. of India, Directorate of Economics and Statistics, Govt. of India, Directorate of Economics and Statistics, Govt. of Jammu and Kashmir and Directorate of Horticulture Planning and Marketing.

3.5.3 Sampling Structure

A multi stage random sampling was adopted for the selection of samples. District, villages and growers (saffron) were first, second, third and fourth stage units respectively. Pulwama district of Kashmir region was selected because the district covers the maximum area under saffron cultivation and production as well. The villages were selected on basis of highest area under saffron cultivation. A total of 95 sample respondents were selected from the District Pulwama. Total samples were distributed by various categories which are presented in table below. Saffron growers constituted 50 (52.63 per cent) of total sample while saffron wholesalers constituted 15 (15.79 per cent), retailers 20 (21.05 per cent) and different agents 10 (10.53 per cent) of total respectively.

3.5.4 List of Selected District (Villages)

One district for the saffron crop falling in Kashmir region of J&K state was selected for this study:

Selected District

(i) Pulwama

Selection of Villages from District Pulwama

S.No.	Crop	Name of Village
1.	SAFFRON	CHANDHARA
2.		BEFIN
3.		BARSOO
4.		LETHPORA
5.		ANDRUSSU
6.		KONIBAL
7.		DUSOO
8.		KADLABAL
9.		NAMLABAL
10.		PAMPORA

3.5.5 Analysis of Data

Some of the important points in the analysis of the data and the methodology adopted has been elucidated below:

3.5.6 Calculation of Growth and Compound Annual Growth Rate

Compound annual growth rate (CAGR) is a business and investing specific term for the geometric progression ratio that provides a constant rate of return over the time period. CAGR is not an accounting term, but it is often used to describe some element of the business, for example revenue, units delivered, registered users, etc. CAGR dampens the effect of volatility of periodic returns that can render arithmetic means irrelevant. It is particularly useful to compare growth rates from different data sets such as revenue growth of companies in the same industry.

$$CAGR(t_0,t_n) = \left(\cdot \frac{V(t_n)}{} V(t_0) \right)^{\frac{1}{t_n - t_0}} - 1$$

where,

$V(t_0)$ = Start value

$V(t_n)$ = Finish value

$t_n - t_0$ = number of years.

The percent change from one period to another is calculated from the formula:

$$PR = \frac{(V_{Present} - V_{Past})}{V_{Past}} \times 100$$

where,

PR = Per cent Rate

$V_{Present}$ = Present or Future Value

V_{Past} = Past or Present Value

The annual percentage growth rate is simply the percent growth divided by N, the number of years.

3.5.7 Marketing Margins, Costs and Loss

The post harvest loss at various stages of marketing has been included either in the farmer's net margin or market intermediaries' margin. In the present study, the marketing loss at different stages has been used for separating the 'post harvest loss during marketing' at different stages of marketing as well as for estimating the producer's share, marketing margins and marketing loss.

3.5.8 Net Farmers Price

The net price received by the farmer will be estimated as the difference in gross price received by him and sum of his marketing costs and value loss during harvesting, grading, transit and marketing. Thus, the net farmer's price is expressed mathematically as follows:

$NP_F = GP_F - \{C_F + (L_F \times GP_F)\}$ or

$$NP_F = \{GP_F\} - \{C_F\} - \{L_F \times GP_F\} \quad (1)$$

Where NP_F is net price received by the farmers (Rs/10gm),

GP_F is gross price received by the farmers or wholesale price to farmers (Rs/10gm),

C_F is the cost incurred by the farmers during marketing (Rs/10gm),

L_F is physical loss in produce from harvest till it reaches assembly market (per10gm).

3.5.9 Marketing Margins

The margins of market intermediaries include profit and return, which accrue to them for storage, the interest on capital and establishment after adjusting for the marketing loss due to handling. The general expression for estimating the margin for intermediaries is given below.

Intermediaries Margin	= Gross price (sale price)	– Price paid (cost price)	– Cost of marketing	– Loss in value during wholesaling

Net marketing margin of the wholesaler is given mathematically by

$MMw = GPw - GPF - Cw - (Lw \times GPw)$ or

$$MMw = \{GPw - GPF\} - \{Cw\} - \{Lw \times GPw\} \quad (2)$$

Where MMw is net margin of the wholesaler (Rs/10gm),

GPw is wholesaler's gross price to retailers or purchase price of retailer (Rs./10gm)

Cw is cost incurred by the wholesalers during marketing (Rs./10gm),

Lw is physical loss in the produce at the wholesale level (per 10gm)

In the marketing chain, when more than one wholesaler is involved, *i.e.*, primary wholesaler, secondary wholesaler, etc, then the total margin of the wholesaler is the sum of the margins of all wholesalers. Mathematically,

$$MMw = MMw1 + ... + MMwi + ... + MMwn$$

Where MMwi is the marketing margin of the i-th wholesaler.

Net marketing margin of retailer is given by:

$$MMR = GPR - GPW - CR - (LR \times GPR) \text{ or}$$

$$MMR = \{GPR - GPW\} - \{CR\} - \{LR \times GPR\} \quad (3)$$

Where MMR is net margin of the retailer (Rs./10gm),

GPR is price at the retail market or purchase price of the consumers (Rs./10gm)

LR is physical loss in the produce at the retail level (per 10gm),

CR is the cost incurred by the retailers during marketing (Rs./10gm).

The first bracketed term in equations (1), (2) and (3) indicates the gross return, while the second and third bracketed terms indicate respectively the cost and loss at different stages of marketing.

Thus, the total marketing margin of the market intermediaries (MM) is calculated as

$$MM = MMW + MMR \quad (4)$$

Similarly, the total marketing cost (MC) incurred by the producer/seller and by various intermediaries is calculated as

$$MC = CF + CW + CR \quad (5)$$

Total loss in the value of produce due to injury/damage caused during handling of produce from the point of harvest till it reaches the consumers is estimated as

$$ML = \{LF \times GPF\} + \{LW \times GPW\} + \{LR \times GPR\} \quad (6)$$

3.5.10 Marketing Efficiency

Most commonly used measures are conventional input to output marketing ratio, Shepherd's ratio of value (price) of goods marketed to the cost of marketing (Shephard, 1965) and Acharya's modified marketing efficiency formula (Acharya and Agarwal, 2001). However, all these measures do not explicitly mention the loss in the produce during the marketing process as a separates item in marketing. As reduction in loss itself is one of the efficiency parameters, there is a need to incorporate this component explicitly in the existing marketing ratios to get correct measures of marketing efficiency while comparing alternate markets/channels. 'Marketing loss' component is incorporated in the widely used formula as given by

Acharya and Agarwal (2001) and the modified marketing efficiency (ME) formula is given below.

$$ME = \frac{NP_F}{MM + MC + ML}$$

where,

NPF is net price received by the farmers (Rs/10gm),

MM is the marketing margin,

MC is marketing cost,

ML is marketing loss.

3.6 Results

The results of the present study are given in this chapter. Saffron is emerging as an important spice crop in india. Though Jammu and Kashmir State is on eof the traditional saffron growing states in india, the productivity and marketing is very low and poor when compared to national and international average therefore, efforts need to be taken to enhance its productivity and strengthen the marketing structure of this spice crop.

3.6.1 Growth Performance of the Saffron Crop

The country wise area, production and productivity were depicted in Table 1.1. From the above cited table, it is observed that Iran has the highest area (43,408 ha) under saffron cultivation followed by India *i.e.*, 4265 ha, then Greece (1000ha) and soon. As far as its production is concerned again India follows Iran (174.00MT). but with regard to productivity, India rank is fifth. Italy, which has lowest area and production, rank first in productivity as is clear from the table.

Table 3.1, provides an insight across the district wise area under saffron cultivation and its production. The above cited table highlights that in Jammu and Kashmir Pulwama appears to be at the highest scale in terms of area (1851.75 ha) and production having 1.85 MT, followed by Budgam 398.28 ha and production having 0.40 MT, while on the other hand Doda was having lowest area and production *i.e.*, 61.73 ha and 0.06MT production.

Table 3.2 highlights the domestic price of saffron in India per kg during 2009-10 to 2014-15. The above cited table highlighted the monthly fluctuations of the domestic price of saffron. It has been identified that during 2009-10, the price ranges from 3.2lakh/kg to 1.7 lakh/kg from April to march. On the similar lines, the year 2013-14 highlights that the price/kg of saffron was lowest from 1.1lakh/kg in April with 1.85lakh per kg as highest during September and October. From the table, it was further observed that during 2014-15, the price of saffron was almost same throughout the year ranging from 1.63lakh/kg in April to 1.67lakh/kg. The domestic price on an average was highest during 2009-10 (2.70 lakh/kg) with lowest (1.11 lakh/kg) during 2011-12.

Table 3.1: District-wise Area Under Saffron Cultivation and its Production

District	*Area (hectares)*	*Production (MT)*	*Productivity (kg/ha)*
Srinagar	157.28	0.16	1.01
Budgam	398.24	0.40	1.00
Pulwama	1851.75	1.85	0.99
Doda	61.73	0.06	0.97

Source: Annual Report of Directorate of Agricultural Jammu and Kashmir.

Table 3.2: Average Domestic Price of Saffron in India

Year	*2009-10*	*2010-11*	*2011-12*	*2012-13*	*2013-14*	*2014-15*
APR	321000	178500	101000	–	111375	163750
MAY	321000	170500	–	–	130000	167500
JUN	301000	152250	–	–	135000	167500
JUL	308500	150950	–	–	141800	167500
AUG	304800	163500	121000	115500	162000	–
SEP	321000	163500	–	115500	185000	–
OCT	319600	151500	–	115500	185000	–
NOV	251000	133500	–	118000	163500	–
DEC	236000	131000	–	125500	160000	–
JAN	203167	131000	–	125500	159500	–
FEB	181375	126000	–	125500	162500	–
MAR	173500	106000	–	125500	162500	–
Average	**270162**	**146517**	**111000**	**120813**	**154848**	**166563**

Source: Spice Board of India.

Compound annual growth rate of saffron in Jammu and Kashmir State in terms of area, production and yield presented in Table 3.3 and Figure 3.1 respectively. The Table 3.3 depicted that on an average, the compound annual growth rate showed a declining trend in terms of area -5.80 per cent, production -3.28 per cent and productivity or yield -0.419 per cent respectively. In contrast to this while analyzing the growth rate over the period of time, there has been increase in area up to 14.62 per cent during 2004-05, afterwards continuous downtrend has been observed except for 2009-10, where the growth trend was highest of 15.4 per cent. On the similar lines the highest production as well as yield rate appears to be 68.52 per cent and 79.53 per cent during the year 2002-03 *i.e.*, 68.52 per cent and 79.53 per cent respectively. It was further observed from the table that the annual growth trend was a mixture of increase and decrease. In some period, it increased, while in another, it decreased as is clear from the table.

Figure 3.2 depicted the size of population of saffron growers in rural and peri-urban area was 65.38 per cent and 34.62 per cent, respectively. Out the total male population of saffron growers, 65.21 per cent were in rural area and 34.79 per cent

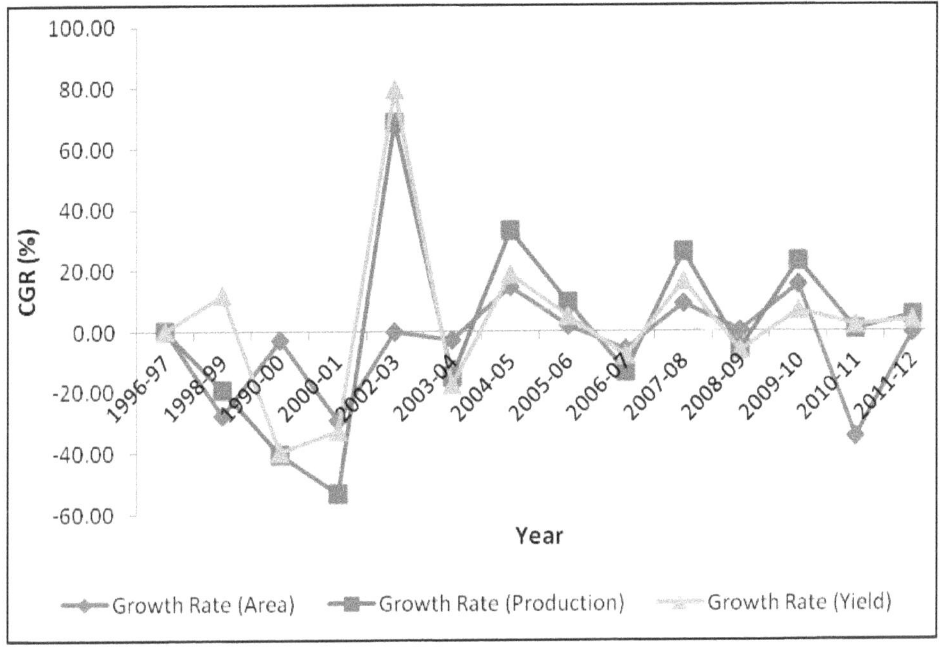

Figure 3.1: Compound Annual Growth Rate for Saffron in J&K State.

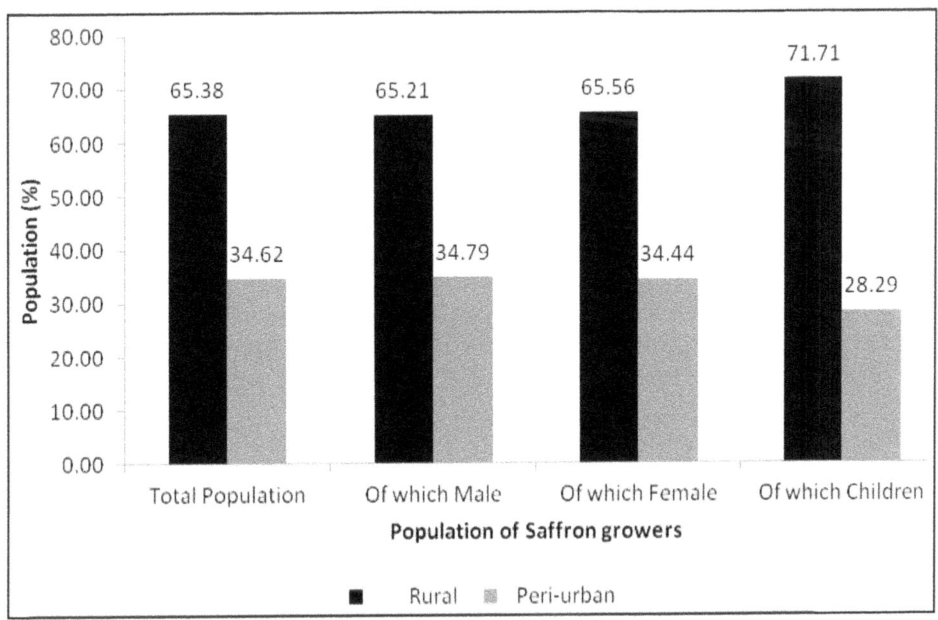

Figure 3.2: Sample Farmers Confined to Saffron Cultivation.

in peri-urban area and female population. It was 65.56 per cent and 34.44 per cent respectively while as in case of children's it was 71.71 per cent and 28.29 per cent respectively.

Table 3.3: Compound annual growth rate for saffron in J&K state

Year	Area (ha)	Production (MT)	Yeild (kg/Ha)	Growth Rate (Area)	Growth Rate (Production)	Growth Rate (Yield)
1996-97	5707	15.95	2.8	0	0	0
1998-99	4116	12.88	3.13	–27.88	–19.25	11.79
1990-00	3997	7.65	1.89	–2.89	–40.61	–39.62
2000-01	2831	3.59	1.27	–29.17	–53.07	–32.8
2002-03	2825	6.05	2.28	–0.21	68.52	79.53
2003-04	2742	5.15	1.88	–2.94	–14.88	–17.54
2004-05	3143	6.86	2.23	14.62	33.2	18.62
2005-06	3200	7.5	2.34	1.81	9.33	4.93
2006-07	3010	6.5	2.15	–5.94	–13.33	–8.12
2007-08	3280	8.2	2.5	8.97	26.15	16.28
2008-09	3280	7.7	2.34	0	–6.1	–6.4
2009-10	3785	9.46	2.5	15.4	22.86	6.84
2010-11	2479.2	9.5	2.55	–34.5	0.42	2
2011-12	2469.02	10	2.64	–0.41	5.26	3.53
CAGR	**–0.058**	**–0.033**	**–0.004**			

The average size of family and landholding of saffron farmers is presented in Table 3.4. The perusal of table depicts that 42 per cent of the saffron farmers with having an average landholding between 2.0 to 5.0 kanals are classified under the category of marginal farmers. Whereas 5 per cent of the farmers with land holding of 8.0 to 9.0 kanals and size of family >12 are classified under the large category. The villages show cultivators to be less than 50 per cent. Family size varies from 3-20 members. The farmers falling in class <4, 5-9 and >10 members are 1 per cent, 91 per cent and 8 per cent respectively. Maximum farmers, *i.e.* 42 per cent have land holdings size of 2.0 to 5.0 kanals are categorized as marginal farmers, followed by 38 per cent having land holding size between 5.5 to 8.0 kanal, thus categorized as small farmers. 15 per cent farmers are categorized as medium farmers with an average land holdings size ranging from 8.0 to 9.0 kanals and 5 per cent farmers categorized as large farmers having 9.0 and above land holding size. The study thus confirmed that the majority of farmers in the saffron villages are marginal, having a low risk bearing capacity. Thus it can be concluded that the farmer under saffron cultivation are residing in rural areas with an average landholding 5.5 to 8.0 kanals and contributing to the total production of 10MT with the export value 1.5 million MT respectively. The maximum annual income per farmer comes from saffron production followed by apple cultivation. On an average, farmer owning an area of about 0.6995 ha per farmer under saffron with an average production of about 0.016732 quintals, earns an annual income of 6300 US$.

Table 3.4: Population, Family Size and Landholding of Saffron Farmers of Sample Area

Class (Size of Family)	*Land Holding (Kanal)*	*Percentage of Farmers*	*Farmers*
< 4 (1 per cent)	2.0-5.0	42 per cent	Marginal
5-9 (89 per cent)	5.5-8.0	38 per cent	Small
> 10 (8 per cent)	8.0-9.0	15 per cent	Medium
>12 (2 per cent)	9.0 and above	5 per cent	Large

3.6.2 Value Chain Activity of Stakeholders

The Table 3.5 revealed the role of different actors in the saffron value chain. The saffron value chain comprises of farmers, wholesalers, agents and retailers. In addition to this government play an important and vital role for smooth functioning of the activities in the saffron value chain. For the smooth functioning of the activities in the value chain the value chain actors play their vital role at their own places at a particular period of time. The above table cited the role and activities performed by the various chain actors for the smooth functioning of the saffron value chain.

Table 3.5: Actors and their Activities in the Saffron Value Chain

Actors	*Processors*								
	IB	*DBTF*	*PB*	*PLFPS*	*HS*	*SSDS*	*BPM*	*WT*	*RT*
Trader	✔								
Government		✔	✔						
Farmers			✔	✔	✔	✔			
Wholesalers			✔	✔	✔	✔	✔	✔	
Retailers			✔	✔	✔	✔	✔		✔

*Note: **IB**: Importing bulbs, **DBTF**: Distributing bulbs to farmers, **PB**: Producing bulbs, **PLFPS**: Preparing land for plantation saffron, **HS**: Harvesting Saffron, **SSDS**: Separation stigma and drying saffron, **BPM**: Branding, packaging and marketing, **WT**: Wholesale trading, **RT**: Retail trading.

3.6.3 Value Chain Models Existed

The agricultural value chain concept is the idea of actors connected along a chain producing and delivering goods to consumers through a sequence of activities. However, this "vertical" chain cannot function in isolation and an important aspect of the value chain approach is that it also considers "horizontal" impacts on the chain, such as input and finance provision, extension support and the general enabling environment. The approach has been found useful, particularly by donors, in that it has resulted in a consideration of all those factors impacting on the ability of farmers to access markets profitably, leading to a broader range of chain interventions. It is used both for upgrading existing chains and for donors to identify market opportunities for small farmers.

The Flow Chart 5.1 revealed that the saffron value chain of channel-I which comprised of value chain actors that perform various value addition activities

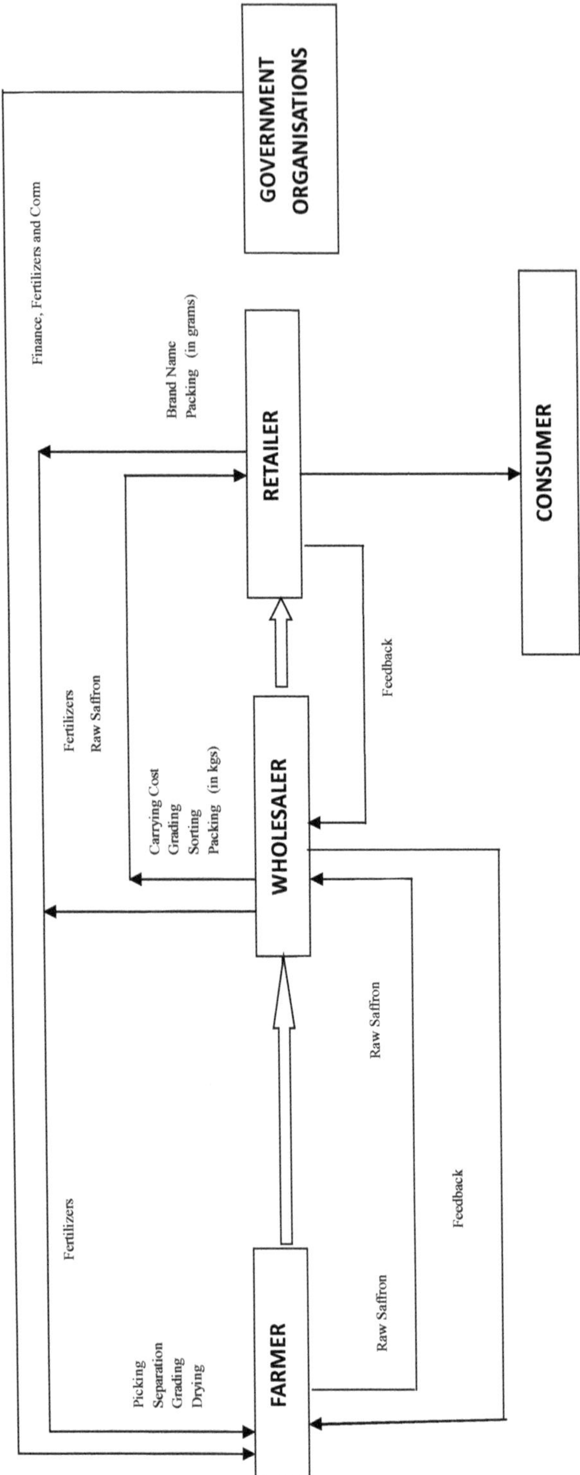

Flow Chart 3.1: Value Chain of Saffron in Kashmir for Channel-1.

during particular period of time. The actors involved in Channel-I were farmers, wholesalers, retailers, government, and consumer. At different stages the product went under different value addition by the channel partners. Farmers were involved in the production activity of saffron (pre and post) harvesting practices and drying activity. Wholesalers were involved in collecting the raw saffron from the Farmers and then incorporate sorting, grading and re-dying it for further business purpose. Retailers were involved in the practice of taking the product from the wholesaler then indulging in the practice of re-grading, branding, and processing of the product. In addition to this government and other financial institutions played a vital role for the effective and efficient production process of the saffron by the farmers at their respective areas. In this channel, the flow of product started and shifted from the farmer to the wholesaler, wholesaler to the retailer and retailer to the end consumer respectively in a value added manner in order to sell the product to the customer with having value added features or to sell the product to the consumer with a value embedded in it for their satisfaction.

The Flow Chart 5.2 revealed that the saffron value chain of channel-II which comprised of value chain actors that performed various value addition activities during particular period of time. The actors involved in Channel-II were farmers, agents, wholesalers, retailers, government, and consumer. At different stages the product went under different value addition by the channel partners. Farmers were involved in the production activity of saffron (pre and post) harvesting practices and drying activity. Agents collected the material in raw form and sorted it in bulk for further sale. Wholesalers were involved in collecting the raw saffron from the farmers/agents and then incorporated sorting, grading and re-dying it for further business purpose. Retailers were involved in the practice of taking the product from the wholesaler then indulging in the practice of re-grading, branding, and processing of the product. In addition to this government and other financial institutions played a vital role for the effective and efficient production process of the saffron by the farmers at their respective areas. In this channel the flow of product started and shifts from the farmer to the agent, agent to the wholesaler, wholesaler to the retailer and retailer to the end consumer respectively in a value added manner in order to sell the product to the customer with having value added features or to sell the product to the consumer with a value embedded in it for their satisfaction.

3.6.4 Economic Relationship (Chain-I and II)

The economic relationship of stakeholders depicts the level of dependency on each other. The Flow Chart 5.3 revealed and depicted the relationship between the various chain actors in terms of channel-I was concerned. *i.e.,* it showed the relationship between the farmer, wholesaler, retailer and consumer with each other in economical terms. From the fig. it was observed that the marketing cost incurred by farmer was Rs 2417.44/10gm, while as the physical loss was to the extent of Rs 448.76/10gm. As far as the wholesaler was concerned, the marketing cost was Rs 2549.60/10gm and for retailer, it was Rs 2602.23/10gm for retailer. The physical loss for wholesaler was Rs 267.17/10gm while as it was Rs 150.27/10gm. The marketing margin of wholesaler was Rs 138.23/10gm and for retailer, it was Rs 1464.31/10gm.

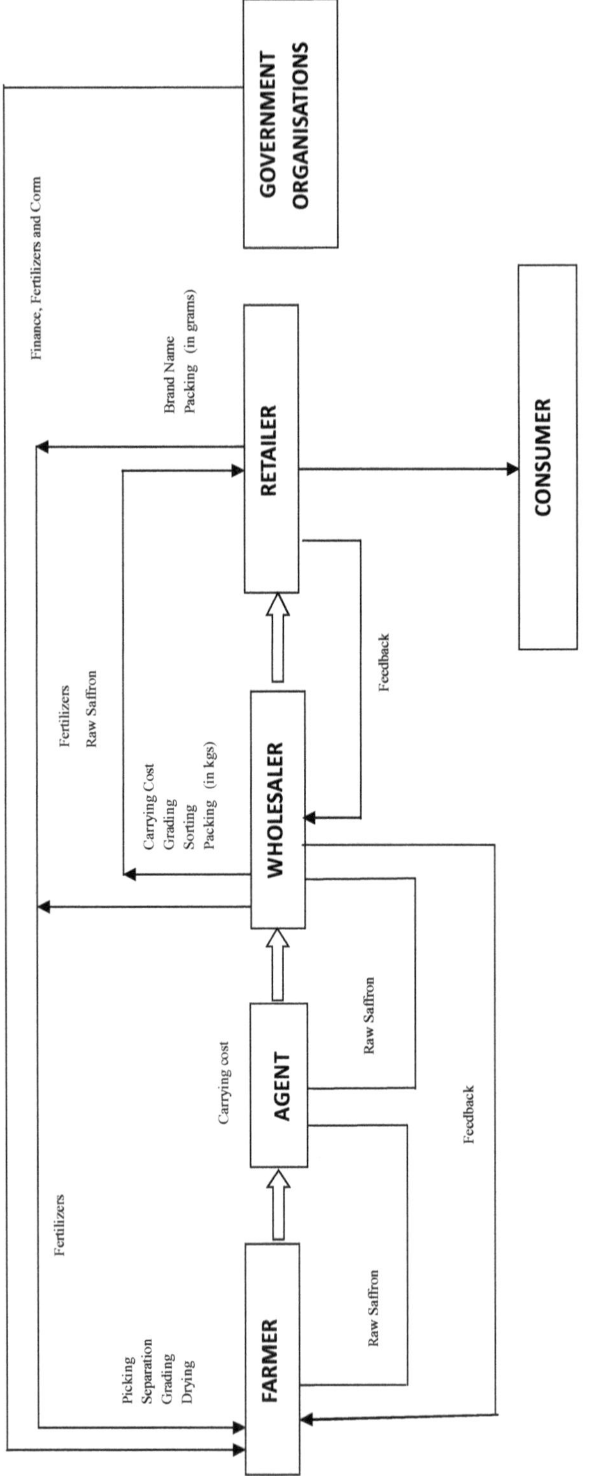

Flow Chart 3.2: Value Chain of Saffron in Kashmir for Channel-II.

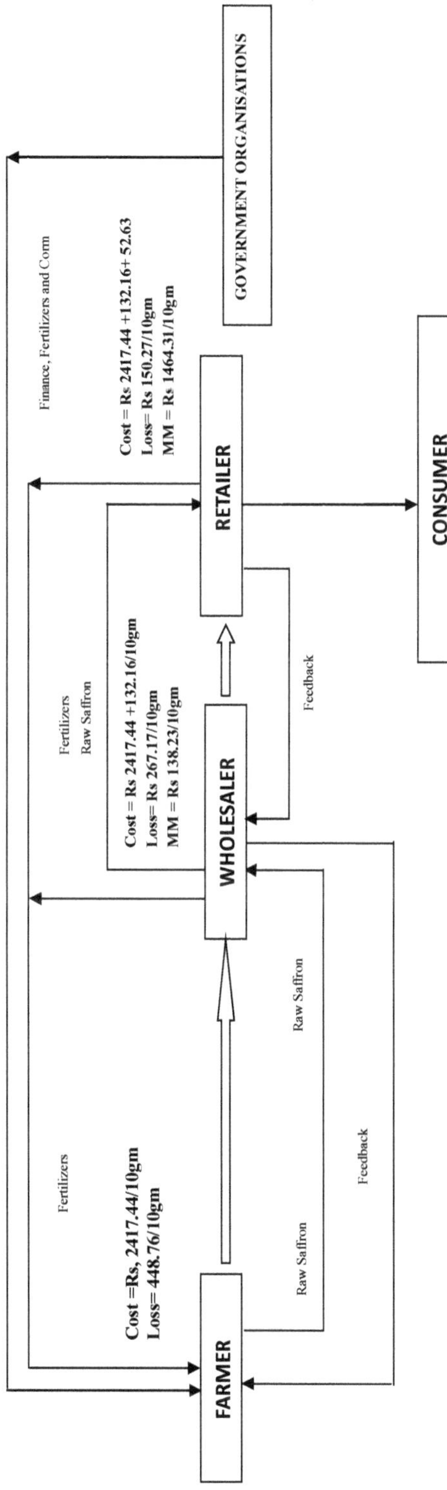

Flow Chart 3.3: Economic Relationship Model for Channel-I.

The Flow Chart 5.4 revealed and depicted the relationship between the various chain actors in terms of channel-II was concerned. *i.e.*, it showed the relationship between the farmer, agent wholesaler, retailer and consumer with each other in economical terms. From the fig. it was observed that the marketing cost incurred by farmer was Rs 2417.44/10gm, while as the physical loss was to the extent of Rs 448.76/10gm. As far as agent was concerned, the marketing cost was Rs 2436.86/10gm and physical loss was having Rs 67.44/10gm with margin Rs 580.18/10gm. As far as the wholesaler was concerned, the marketing cost was Rs 2549.60/10gm and for retailer, it was Rs 2602.23/10gm. The physical loss for wholesaler was Rs 267.17/10gm while as it was Rs 150.27/10gm for retailer. The marketing margin of wholesaler was Rs 138.23/10gm and for retailer, it was Rs 1464.31/10gm.

3.7 Financial and Governmental Support

The Figure 3.3 revealed that the in saffron sample area, the main financing bodies were Jammu and Kashmir Bank, Punjab National Bank, State Bank of India and Ellaqui Dehati Bank. The main source of financing was in shape of Kissan Credit Cards. The figure depicted that out of the total financing disbursement the role of Jammu and Kashmir Bank Limited was more 65 per cent followed by 15 per

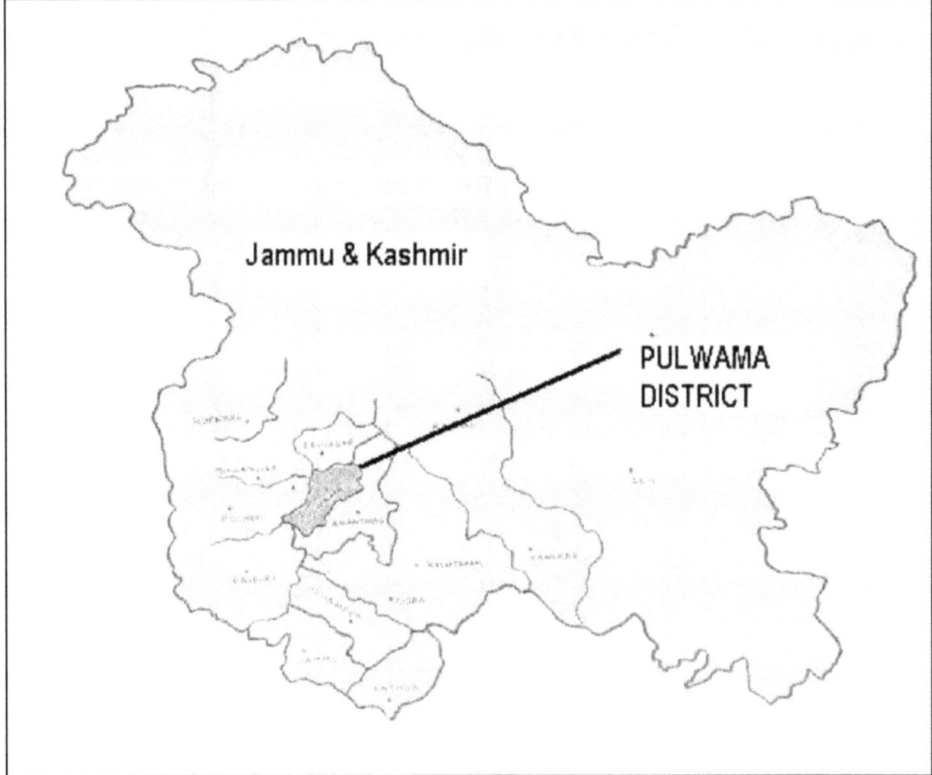

Figure 3.3: Map of Sample Area.

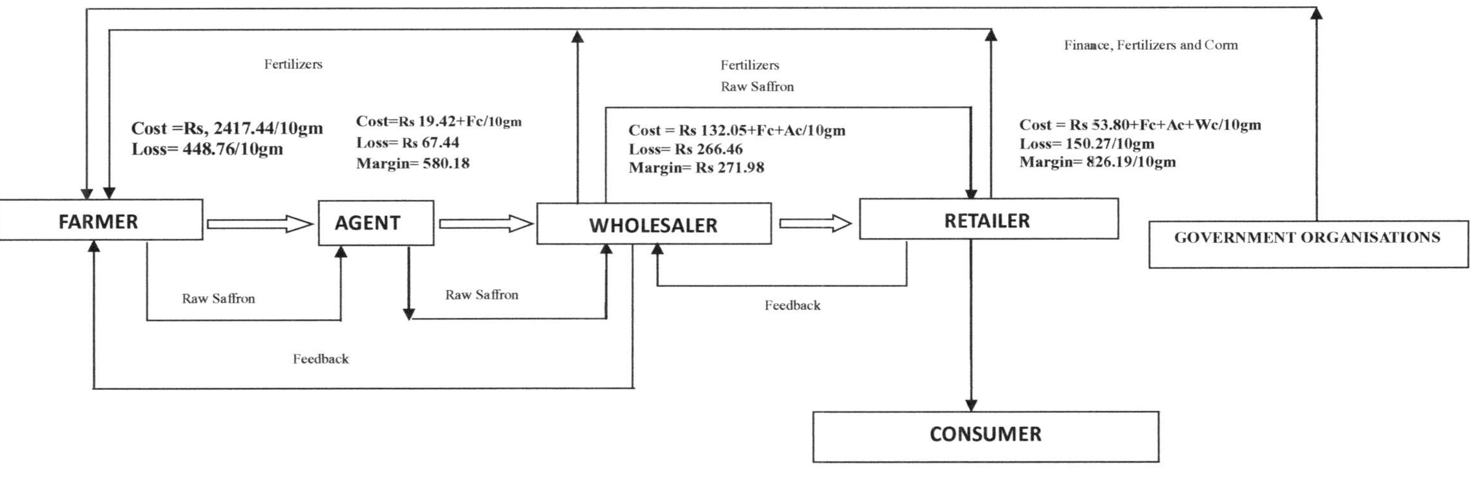

Flow Chart 3.4: Economic Relationship Model for Channel-II.

***Note*: Fc: Farmers incurred marketing cost; Ac: Agents incurred marketing cost and WC: Wholesalers incurred marketing cost.**

cent each by State Bank of India and Ellaqui Dehati Bank and 5 per cent by Punjab National Bank. Among the saffron farmers the medium of finance is maximum through the Kissan Credit Cards.

Agricultural value chain finance is concerned with the flows of funds to and within a value chain to meet the needs of chain actors for finance, to secure sales, to buy inputs or produce, or to improve efficiency. Examining the potential for value chain finance involves a holistic approach to analyze the chain, those working in it, and their inter-linkages. These linkages allow financing to flow through the chain. For example, inputs can be provided to farmers and the cost can be repaid directly when the product is delivered, without need for farmers taking a loan from a bank or similar institution. This is common under contract farming arrangements. Types of value chain finance include product financing through trader and input supplier credit or credit supplied by a marketing company or a lead firm. Other trade finance instruments include receivables financing where the bank advances funds against an assignment of future receivables from the buyer, and factoring in which a business sells its accounts receivable at a discount. Also falling under value chain finance is asset collateralization, such as forward contracting, futures and insurance.

According to the various estimates, total turnover of the saffron crop is around Rs 400 crore per year. Therefore financing of the saffron (cultivation and marketing) becomes a central issue. It includes saffron financing scheme and KCC cards.

Financial institutions involved in this activity includes: - J&K Bank, Elaqui Dehati Bank and State Bank of India through which the saffron growers are taking the financial assistance for fulfilling their need during the saffron season. The Practices, Schemes/modes are as follows:

3.7.1 Zafran Finance

To provide adequate and need based financial assistance for cultivation of saffron. The term loan covers the entire plantation and production costs including plant material, agricultural machinery, labour, etc. Post-harvest and Packaging costs are also covered.

3.7.1.1 Nature of Facility

Agricultural Term Loan.

3.7.1.2 Eligibility

All saffron growers including small, marginal and large farmers including contract farmers engaged in cultivation of saffron or intending to commence the cultivation of saffron.

3.7.1.3 Minimum and Maximum Amount of Loan

Minimum = Rs 0.06 Lacs

Maximum = Rs 10.00 Lacs.

Post-harvest and packaging costs are also covered under the scheme. All saffron growers including small, marginal and large farmers including contract farmers engaged in cultivation of saffron or intending to commence the cultivation of saffron are eligible to obtain loan under the scheme. For land holding of 5 kanals, loan of 50,000 is granted to a grower. For 10 kanals, loan of Rs 85,000 shall be granted. For 15 kanals and 20 kanals (1 hectare), an amount of Rs.1.19 lacs and Rs.1.54 lacs is granted to a grower under the scheme respectively. The loan facility sanctioned under the scheme shall be disbursed in 2 installments. First Installment constitutes 60 per cent of the loan amount sanctioned and shall be disbursed during year 1(March –September). Remaining 40 per cent of the loan sanctioned (second installment) shall be released in the shape of during year 2 (March –September). However, additional finance of Rs 10,000 per acre for packaging and other related expenditures shall be disbursed at the time of harvest for every productive year during the term of loan and shall not be part of the disbursement schedule. The loan is repayable in equated monthly installments (EMI) within 4 years from the date of first disbursement with an initial moratorium of 6 months.

The growers seeking support of this facility shall have to pay a processing charge of Rs 400/-to be paid upfront in case of third party guarantee and Rs 500/- in case of mortgage. The facility is carrying 13 per cent fixed rate of interest.

3.7.2 Kissan Credit Card

3.7.2.1 J&K Bank Ltd.

With the aim to make farmers financially sound so that they can easily and hassle free does practice of saffron cultivation. The bank provides the saffron farmer with the benefit and scheme of KCC.

3.7.2.1 Purpose

For meeting the cultivation needs of farmers and other short term requirements including those of subsidiary/allied activities and consumption needs.

3.7.2.2 Eligibility

Individual farmers/joint borrowers/partnership firms/private and public Ltd. companies who are owner cultivators/engaged in allied practice.

3.7.2.3 Subsidiary Activities

Farmers who are cultivating authorized leased lands are also eligible. The farmer should not be a defaulter of any financial institution.

3.7.2.4 Loan Amount

Rs 15000/Kanal with ROI @ 4 per cent per annum (conditions apply).

3.7.2.5 Repayment

Credit Limit for three years (with annual and half yearly sub-limits)

3.7.3 Ellaqui Dehati Bank

The bank provides financial support to the saffron farmers in shape of KCC. Till to date only nine cases of saffron farmers for KCC has been registered with the bank as per study.

3.7.3.1 Purpose

For meeting the cultivation needs of farmers and other short term requirements including those of subsidiary/allied activities and consumption needs.

3.7.3.2 Eligibility

Individual farmers/partnership firms/private and public Ltd. companies who are owner cultivators/engaged in allied activities.

3.7.3.3 Subsidiary Activities

Farmers who are cultivating authorized leased lands are also eligible. The farmer should not be a defaulter of any Financial Institution.

3.7.3.4 Loan Amount

15000 per kanal to the saffron farmers with ROI @ 7.5 per cent (Upto 100000 @ 4 per cent) and (Above 100000 @ 7.5 plus)

3.7.4 State Bank of India

This bank provides the agriculture loan facility to the farmers in shape of long and short term loans.

3.7.5 Insurance Coverage Plan (Governmental Support)

In a significant development, government has decided to bring saffron crop in Kashmir under insurance cover in a phased manner. In this connection Agriculture department Kashmir will be implementing Weather-based Crop Insurance Scheme in Kashmir division. In the first instance the saffron crop spread over an area of 2469.02 hectares shall be covered under the scheme. The insurance cover shall be provided at Rs 2 lakh per hectare and the crop shall be covered for deficit rainfall, consecutive dry spell, low temperature and high rainfall. As many as 50 per cent of the premium shall be borne by the government under the scheme, also planned out that Insurance companies shall install an automatic weather station at Pampore for capturing of data on the basis of which the crop shall be covered.

3.7.5.1 Salient Features

All farmers having sanctioned credit limit from financial institutions *i.e.*, Commercial Banks, Regional Rural Banks, and Cooperative Banks/PACS for growing saffron crop in the notified areas are eligible for coverage on compulsory basis.

3.7.5.2 Sum Insured

Rs 2, 00,000/per hectare

3.7.5.3 Premium

@12 per cent of sum insured

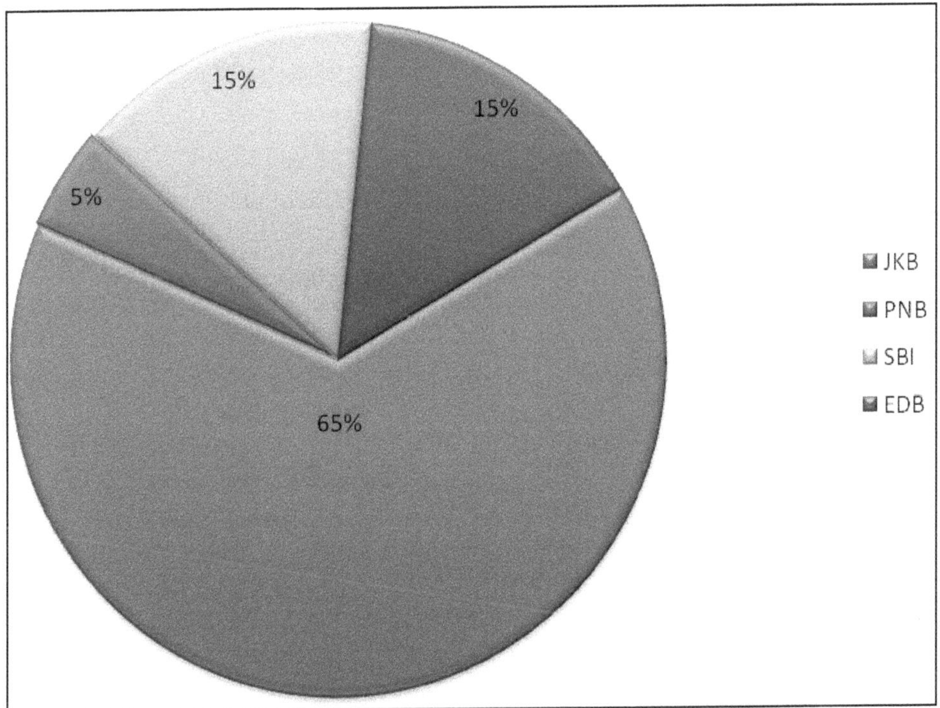

Figure 3.4: Percentage of Kissan Credit Finance Disbursement of different Banks in Sample Area.

Note: Jammu and Kashmir Bank; Punjab National Bank; State Bank of India; Ellaqui Dehati Bank.

3.7.5.4 Subsidy

50 per cent subsidy is available to the farmers.

3.7.5.5 Premium Payable by Farmers

Rs 12,000/per hectare (*i.e.* @6 per cent of sum insured).

3.8 Opportunities and Threats in Saffron chain

The opportunities and threats in saffron chain have been collected from the secondary source of information (Nehvi *et al.*, 2013). Table 3.6 adopted from the book "A review of R&D intervention "Saffron –the Gold Condiment by Nehvi *et al.*, indicated that the saffron cultivation is having a lot of opportunities in terms of expanding international trade, rising demand, expansion of saffron in non-traditional areas, farmers receptivity, government support in terms of finance and input providing for the saffron production purpose, employment generation, control on mal practices and government policies respectively.

However the major threat as per the secondary source of information has been characterized as under:

3.8.1 Crop Diversification

Low return and market volatilities is forcing farmers to substitute saffron cultivation with horticulture crops.

3.8.2 Urbanization of Saffron Belt

Establishment of cement factories and other residential houses on the prime saffron land is becoming a serious concern as far as area under saffron is concerned.

Table 3.6: Opportunities and Threats in Saffron value chain

Opportunities	*Threats*
Expanding international trade	Crop diversification
Rising demand	Urbanization of saffron belt
Expansion of saffron in non-traditional areas	Adulteration
Farmers receptivity	Urbanization of saffron belts
Governmental support	Unavailability of proper timed inputs
Employment Generation	Unfair trade practices
Control on Mal practices	Jeopardizing of stakeholders
Government Policies	Unorganized markets

Source: Secondary Information.

3.8.3 Adulteration

It is a big threat to the saffron industry presently as pure saffron is mixed with the adultered mixtures of various items in order to sell the adultered saffron at high prices in the market.

3.8.4 Unavailability of Inputs

Insufficient supply and limited availability of inputs for the production process affects the productivity in whole. Hence hampers the income level of the saffron farmer.

3.8.5 Unfair Trade Practices

An unorganized market leads to the unfairity of the trade aspect regarding the saffron business.

3.8.6 Unorganized Markets

Lack of organized regulated markets is a great threat to the saffron industry. This leads to the indiscrimination to the saffron farmers in terms of proper return to their produce. Hence decreases the economical level of saffron farmer's large extent.

While as opportunities falling under saffron chain are as under:-expanding international trade, rising demand, expansion of saffron in non-traditional areas, farmers receptivity, governmental support, employment generation, control on mal practice and governmental policies respectively in order to boost the saffron industry at large extent.

3.9 Marketing of Saffron

In planned economic development programme, exchange of goods play a very important role in maintaining equilibrium between production and consumption. The marketing of agriculture products is becoming more important as created by the new world trade order under the WTO agreements. The small and resource poor cultivators have to face big world players for marketing their produce. The prosperity of the cultivators thus depends not only on the increased rate of production, but also upon the methods and efficiency with which they dispose of their produce to the great advantage.

3.9.1 Market Structure of Saffron Marketing

It was imperative to study the activities of each agency taking part in saffron trade in the study or sample area. The agencies that facilitated the flow of saffron spice till they reached the ultimate consumer were commission agents, wholesalers and retailers.

3.9.1.1 Commission Agents

The function of the agent is to sell the product of a producer without any risk of loss and with low cost in the study. In lieu of his service, he charges certain percentage on the total sale value of the commodity.

3.9.1.2 Wholesaler

Wholesalers have got the key position in saffron marketing and also play sometimes the role of agents as well as broker. Generally, they purchased the produce either from the agent or directly from the grower or farmer in the market.

3.9.1.3 Retailer

The retailer is a marketing functionary, who caters the needs of the customers or consumers by retailing, generally keep a small establishments and reap a maximum profit especially in saffron trade due to their higher share in consumers' rupee. Mostly they buy the produce from the wholesaler as well as producers. Owing to these facts survey of the selected farmers in marketing their produce was conducted and the results obtained are presented under various headings:

3.9.2 Marketing Channels in Selected Study Area

The chain of various intermediaries/functionaries commonly known as marketing channel comprising of agencies like producers, commission agents, wholesalers and retailers and also sometimes direct sale of produce help in distribution of the fruits/crops from producer to ultimate consumers in Kashmir region. The marketing channel operating under Kashmir conditions are as under

1. Producer/Farmer → Wholesaler → Retailer → Consumer
2. Producer/Farmer → Commission agents → Wholesaler → Retailer → Consumer

Overview of Marketing of Saffron Spice in Sample Area.

3.9.2.1 Marketing Cost, Margins, Losses and Price Spread

3.9.2.2 Marketing Cost Incurred by the Chain for Channel-I

The Table 3.7 revealed that the major item of marketing expenses in all channels at producer's level included picking, filling, and transportation cost. The average total marketing cost was recorded as Rs. 255.88, Rs. 262.22, Rs. 241.45, Rs Rs. 243.18, Rs. 264.75, Rs. 267.04, Rs. 272.30, Rs. 272.00, Rs. 254.67 and Rs. 268.75 per 10gm respectively for channel-I. for Chandhara, Befin, Barsoo, Letpora, Andrussu, Konibal, Dusoo, Kadlabal, Namlabal, Pampore. Whereas it was Rs. 237.00, Rs. 243.02, Rs. 224.65, Rs. 224.18, Rs. 246.25, Rs. 249.04, Rs. 252.30, Rs. 255.00, Rs. 238.00 and Rs. 248.00 per 10gm respectively for channel-I for the farmer. Marketing cost borne by the wholesaler Rs. 13.38, Rs. 13.60, Rs. 12.80, Rs. 13.25, Rs. 13.25, Rs. 12.80, Rs. 14.33, Rs. 12.33, Rs. 11.67 and Rs. 14.75 per 10gm respectively for channel-I and marketing cost borne by the retailers Rs. 5.50, Rs. 5.60, Rs. 4.00, Rs. 5.75, Rs. 5.25, Rs. 5.20, Rs. 5.67, Rs. 4.67, Rs. 5.00 and Rs. 6.00 per 10gm respectively.

Table 3.7: Marketing Cost Incurred by the Chain for Channel-I (Rs./10gm)

Name of Village	*Marketing Cost of Farmer*	*Marketing Cost of Wholesaler*	*Marketing Cost of Retailer*	*Total Marketing Cost*
Chandhara	237.00	13.38	5.50	255.88
Befin	243.02	13.60	5.60	262.22
Barsoo	224.65	12.80	4.00	241.45
Lethpora	224.18	13.25	5.75	243.18
Andrussu	246.25	13.25	5.25	264.75
Konibal	249.04	12.80	5.20	267.04
Dusoo	252.30	14.33	5.67	272.30
Kadlabal	255.00	12.33	4.67	272.00
Namlabal	238.00	11.67	5.00	254.67
Pampora	248.00	14.75	6.00	268.75

3.9.2.3 Marketing Cost Incurred by the Chain for Channel-II

The Table 3.8 revealed that the major item of marketing expenses in all channels at producer's level included picking, filling, and transportation cost. The average total marketing cost was recorded as Rs. 257.81, Rs. 264.32, Rs. 244.45, Rs. 245.31, Rs. 266.63, Rs. 268.94, Rs. 274.47, Rs. 274.67, Rs. 256.33 and Rs. 269.79 per 10gm respectively for channel-II for Chandhara, Befin, Barsoo, Letpora, Andrussu, Konibal, Dusoo, Kadlabal, Namlabal, Pampore. Whereas it was Rs. 237.00, Rs. 243.02, Rs. 224.65, Rs. 224.18, Rs. 246.25, Rs. 249.04, Rs. 252.30, Rs. 255.00, Rs. 238.00 and Rs. 248.00 per 10gm respectively for channel-II for the farmer. Marketing cost borne by the Agents Rs. 1.94, Rs. 2.10, Rs. 1.80, Rs. 2.13, Rs. 1.88, Rs. 1.90, Rs. 2.17, Rs. 1.92, Rs. 1.67 and Rs. 1.93 per 10gm respectively for channel-II. Marketing cost borne by the wholesaler Rs. 13.38, Rs. 13.60, Rs. 12.80, Rs. 13.25, Rs. 13.25, Rs. 12.80, Rs. 14.33, Rs. 12.83, Rs. 11.67 and Rs. 14.14 per 10gm respectively for channel-II marketing cost borne by the retailers Rs. 5.50, Rs. 5.60, Rs. 5.20, Rs. 5.75, Rs. 5.25, Rs. 5.20, Rs. 5.67, Rs. 4.92, Rs. 5.00 and Rs. 5.71per 10gm respectively.

Table 3.8: Marketing Cost Incurred by the Chain for Channel-II (Rs./10gm)

Name of Village Farmer	*Marketing Cost of Agent*	*Marketing Cost of Wholesaler*	*Marketing Cost of Retailer*	*Marketing Cost of Cost*	*Total Marketing*
Chandhara	237.00	1.94	13.38	5.50	257.81
Befin	243.02	2.10	13.60	5.60	264.32
Barsoo	224.65	1.80	12.80	5.20	244.45
Lethpora	224.18	2.13	13.25	5.75	245.31
Andrussu	246.25	1.88	13.25	5.25	266.63
Konibal	249.04	1.90	12.80	5.20	268.94
Dusoo	252.30	2.17	14.33	5.67	274.47
Kadlabal	255.00	1.92	12.83	4.92	274.67
Namlabal	238.00	1.67	11.67	5.00	256.33
Pampora	248.00	1.93	14.14	5.71	269.79

3.9.2.4 Physical Loss Incurred by the Chain for Channel-I and Channel-II

In this study an attempt has been made to estimate physical losses of saffron at different stages of marketing. But during the survey it was found that marketing losses of saffron were high at producer's level. However Tables 3.9 and 3.10 revealed that the physical losses ranged between Rs. 44.71, Rs. 45.12, Rs. 44.49, Rs. 45.02, Rs. 45.06, Rs. 44.56, Rs. 44.80, Rs. 45.03, Rs. 44.98, Rs. 45.00 per 10gram at Farmers level in case of Chandhara, Befin, Barsoo, Letpora, Andrussu, Konibal, Dusoo, Kadlabal, Namlabal, Pampore respectively followed by wholesaler level as Rs. 26.55, Rs. 26.80, Rs. 26.38, Rs. 26.70, Rs. 26.78, Rs. 26.40, Rs. 26.60, Rs. 26.77, Rs. 27.10, Rs. 27.10 at

Table 3.9: Physical Loss Incurred by the Chain for Channel-I (Rs./10gm)

Name of Village	*Physical Loss of Wholesaler*	*Physical Loss of Retailer*	*Physical Loss of Farmer*	*Total Loss*
Chandhara	26.55	14.79	44.71	86.06
Befin	26.80	15.23	45.12	87.15
Barsoo	26.38	14.88	44.49	85.75
Lethpora	26.70	15.11	45.02	86.83
Andrussu	26.78	15.09	45.06	86.93
Konibal	26.40	14.88	44.56	85.84
Dusoo	26.60	14.98	44.80	86.38
Kadlabal	26.77	15.03	45.03	86.83
Namlabal	27.10	15.27	44.98	87.34
Pampora	27.10	15.00	45.00	87.10

respectively then by retailer level as Rs. 14.79, Rs. 15.23, Rs. 14.88, Rs. 15.11, Rs. 15.09, Rs. 14.88, Rs. 14.98, Rs. 15.03, Rs. 15.27, Rs. 15.00 per 10 gram in channel-I respectively. It was least physical loss at agent level which ranged between Rs 6.70 to Rs 6.84. On the whole, the physical loss ranged between Rs 92.41 in Barsoo to Rs 93.91 in Befin.

Table 3.10: Physical Loss Incurred by the Chain for Channel-II (Rs./10gm)

Name of Village	*Physical Loss of Farmer*	*Physical Loss of Agent*	*Physical Loss of Wholesaler*	*Physical Loss of Retailer*	*Total Loss*
Chandhara	44.71	6.70	26.55	14.79	92.76
Befin	45.12	6.77	26.80	15.23	93.91
Barsoo	44.49	6.66	26.38	14.88	92.41
Lethpora	45.02	6.74	26.75	15.11	93.62
Andrussu	45.06	6.76	26.78	15.09	93.68
Konibal	44.56	6.67	26.40	14.88	92.50
Dusoo	44.80	6.72	26.60	14.98	93.10
Kadlabal	45.03	6.76	26.77	15.03	93.59
Namlabal	44.98	6.84	26.70	15.27	93.78
Pampora	45.00	6.84	26.74	15.00	93.58

3.9.2.5 Marketing Margins for the Channel-I and for Channel-II

Table 3.11 revealed and estimated the marketing margins of saffron chain actors in channel-I and II respectively. In the channel-I the marketing margins at wholesalers level is estimated as Rs. 10.08, Rs. 10.60, Rs. 8.82, Rs. 8.80, Rs. 11.23, Rs. 7.80, Rs. 9.07, Rs. 12.57, Rs. 31.23 and Rs. 28.04 in in Chandhara, Befin, Barsoo, Letpora, Andrussu, Konibal, Dusoo, Kadlabal, Namlabal and Pampore villages of sample area respectively.At retailer level it was calculated as Rs. 131.58, Rs. 162.17, Rs. 148.92, Rs. 155.39, Rs. 149.66, Rs. 147.92, Rs. 147.68, Rs. 145.30, Rs. 151.40 and Rs. 124.29 in Chandhara, Befin, Barsoo, Letpora, Andrussu, Konibal, Dusoo, Kadlabal, Namlabal and Pampore villages of sample area respectively. It was estimated at Agents level as Rs. 54.36, Rs. 55.14, Rs. 52.54, Rs. 52.89, Rs. 55.62, Rs. 51.44, Rs. 54.12, Rs. 56.08, Rs. 74.49 and Rs. 73.52 in Chandhara, Befin, Barsoo, Letpora, Andrussu, Konibal, Dusoo, Kadlabal, Namlabal and Pampore villages of sample area. At wholesaler's level was estimated as Rs. 27.08, Rs. 26.60, Rs. 27.82, Rs. 27.00, Rs. 26.98, Rs. 27.80, Rs. 26.07, Rs. 27.90, Rs. 28.63 and Rs. 26.11 in Chandhara, Befin, Barsoo, Letpora, Andrussu, Konibal, Dusoo, Kadlabal, Namlabal and Pampore villages of sample area.At retailers end, it was estimated as Rs. 82.71, Rs. 82.17, Rs. Rs 82.92, Rs. 82.14, Rs. 82.66, Rs. 82.92, Rs. 82.35, Rs. 83.30, Rs. 82.73 and Rs. 82.29 in Chandhara, Befin, Barsoo, Letpora, Andrussu, Konibal, Dusoo, Kadlabal, Namlabal and Pampore villages of sample area respectively per 10 gram. It was thus concluded that the Channel-I having less marketing margin as compared to Channel-II.

Table 3.11: Marketing Margin for Channel-I and II Respectively

Margins	*Name Village*									
	Chandhara	*Befin*	*Barsoo*	*Lethpora*	*Andrussu*	*Konibal*	*Dusoo*	*Kadlabal*	*Namlabal*	*Pampore*
					(Channel–I)					
MMw	10.08	10.6	8.82	8.8	11.23	7.8	9.07	12.57	31.23	28.04
MMR	131.58	162.2	148.92	155.39	149.66	147.92	147.7	145.3	151.4	124.29
					(Channel–II)					
MMA	54.36	55.14	52.54	52.89	55.62	51.44	54.12	56.08	74.49	73.52
MMw	27.08	26.6	27.82	27	26.98	27.8	26.07	27.9	28.63	26.11
MMR	82.71	82.17	82.92	82.14	82.66	82.92	82.35	83.3	82.73	82.29

Note: MMw: Mrketing margin of wholesaler; **MMR**: Marketing margin of retailer; **MMA**: Marketing margin of agent.

3.9.2.6 Price Spread Analysis for Channel-I and Channel-II

The price spread is the gap between the price paid by the consumer and the price received by the farmer at a particular time because from the producer, it has to pass through various agencies before it reaches to the final consumer. The distribution channel in addition to some charges for the services, they perform some margin money is also expected by them in that transaction. Therefore it is worthwhile to examine as to what share of the rupee paid by the consumer was received by the producer. The price spread as percent of consumer rupee for different market functionaries of saffron under different channels of District Pulwama in Kashmir of J&K state was presented in Tables 3.12 and 3.13 respectively. The saffron growers of pulwama in Kashmir region received the net price of about Rs. 996, Rs. 1001, Rs. 1002, Rs. 1017, Rs. 996, Rs. 979, Rs. 983, Rs. 987, Rs. 1002, and Rs. 993 per 10gm for channel-I and channel-II in Chandhara, Befin, Barsoo, Letpora, Andrussu, Konibal, Dusoo, Kadlabal, Namlabal and Pampore villages of sample area respectively. The producer sale price of saffron was Rs. 1277.50, Rs. 1289.00, Rs. 1271.00, Rs. 1286.25, Rs. 1287.50, Rs. 1273.00, Rs. 1280.00, Rs. 1285.00, and Rs. 1285.71 per 10gm in channel-I and II in Chandhara, Befin, Barsoo, Letpora, Andrussu, Konibal, Dusoo, Kadlabal, Namlabal and Pampore villages of sample area respectively respectively. Thus the table revealed that the producer's share in consumer's rupee in Channel-I was Rs. 0.67, Rs. 0.66, Rs. 0.67, Rs. 0.67, Rs. 0.66, Rs. 0.66, Rs. 0.66, Rs. 0.66, Rs. 0.66, Rs. 0.66 per 10 gram respectively in Chandhara, Befin, Barsoo, Letpora, Andrussu, Konibal, Dusoo, Kadlabal, Namlabal and Pampore villages of sample area and in channel-II was Rs. 0.65, Rs. 0.65, Rs. 0.66, Rs. 0.65, Rs. 0.64, Rs. 0.64, Rs. 0.64, Rs. 0.64, Rs. 0.65, Rs. 0.64 per 10 gram respectively in Chandhara, Befin, Barsoo, Letpora, Andrussu, Konibal, Dusoo, Kadlabal, Namlabal and Pampore villages of sample area. Hence showed that the channel-I is more beneficial for a farmer to dispose off their produce in the sample area I order to get a remunerative fair income for their produce.

3.9.2.7 Marketing Efficiency in different Channels

The marketing efficiency is an important tool and therefore requires more attention. The marketing efficiency of different marketing channels of pulwama district in Kashmir of J&K state is shown in Table 3.14 and Figure 3.5 respectively. The saffron growers received highest net return/10gm from channel-I followed by channel-II, whereas the marketing cost/10gm was found to be highest in channel-II followed by Channel-I. The table revealed that the marketing efficiency of Channel-I was recorded as 4.93, 4.28, 4.62, 4.52, 4.50, 4.56, 4.50, 4.58, 4.15, 4.64 per 10 gram in Chandhara, Befin, Barsoo, Letpora, Andrussu, Konibal, Dusoo, Kadlabal, Namlabal and Pampore villages of sample area and 4.27, 4.28, 4.34, 4.39, 4.25, 4.26, 4.22, 4.20, 4.96, 4.94 per 10 gram Chandhara, Befin, Barsoo, Letpora, Andrussu, Konibal, Dusoo, Kadlabal, Namlabal and Pampore villages of sample area respectively. Hence channel-I was found to be the most efficient marketing channel for the grower of saffron followed by channel-II.

Table 3.12: Price Spread Analysis for Channel-I (Rs./10gm)

Name of Village	*Chandhara*	*Befin*	*Barsoo*	*Lethpora*	*Andrussu*	*Konibal*	*Dusoo*	*Kadlabal*	*Namlabal*	*Pampore*
Gross farmers price	1277.5	1289	1271	1286.25	1287.5	1273	1280	1286.67	1285	1285.71
Cost incurred by farmer	237	243.02	224.65	224.18	246.25	249.04	252.3	255	238	247.23
Gross price of wholesaler	1327.5	1340	1319	1335	1338.75	1320	1330	1338.33	1355	1355
Gross price of retailer	1479.38	1523	1488	1511.25	1508.75	1488	1498.33	1503.33	1526.67	1500
Cost incurred by wholesaler	13.38	13.6	12.8	13.25	13.25	12.8	14.33	12.33	11.67	14.14
Cost incurred by retailer	5.5	5.6	5.2	5.75	5.25	5.2	5.67	4.67	5	5.71
Loss at wholesale level	26.55	26.8	26.38	26.7	26.78	26.4	26.6	26.77	27.1	27.1
Loss at retailer level	14.79	15.23	14.88	15.11	15.09	14.88	14.98	15.03	15.27	15
Net farmers price	995.79	1000.87	1001.87	1017.05	996.19	979.41	982.9	986.63	1002.03	993.49
Marketing margin of wholesaler	10.08	10.6	8.82	8.8	11.23	7.8	9.07	12.57	31.23	28.04
Marketing margin of retailer	131.58	162.17	148.92	155.39	149.66	147.92	147.68	145.3	151.4	124.29
Price paid by Consumer	1479.38	1523	1488	1511.25	1508.75	1488	1498.33	1503.33	1526.67	1500
Producer Share in Consumer's Rupee	0.67	0.66	0.67	0.67	0.66	0.66	0.66	0.66	0.66	0.66

Table 3.13: Price Spread Analysis for Channel-II (Rs./10gm)

Name of Village	*Chandhara*	*Befin*	*Barsoo*	*Lethpora*	*Andrussu*	*Konibal*	*Dusoo*	*Kadlabal*	*Namlabal*	*Pampore*
Gross farmers price	1277.5	1289	1271	1286.25	1287.5	1273	1280	1286.67	1285	1285.71
Net farmers price	996	1001	1002	1017	996	979	983	987	1002	993
Farmer Sell to Agent	1277.5	1289	1271	1286.25	1287.5	1273	1280	1286.67	1285	1285.71
Agent to Wholesaler	1340.5	1353	1332	1348	1351.75	1333	1343	1351.33	1368	1368
Gross price of wholesaler	1407.5	1420	1399	1415	1418.75	1400	1410	1418.33	1435	1435
Gross price of retailer	1510.5	1523	1502	1518	1521.75	1503	1513	1521.33	1538	1538
Cost incurred at agent	1.94	2.1	1.8	2.13	1.88	1.9	2.17	1.83	1.67	1.93
Cost incurred at wholesaler	13.38	13.6	12.8	13.25	13.25	12.8	14.33	12.33	11.67	14.14
Cost incurred at retailer	5.5	5.6	5.2	5.75	5.25	5.2	5.67	4.67	5	5.71
Loss at agent level	6.7	6.77	6.66	6.74	6.76	6.67	6.72	6.76	6.84	6.84
Loss at wholesale level	26.55	26.8	26.38	26.75	26.78	26.4	26.6	26.77	26.7	26.74
Loss at retailer level	14.79	15.23	14.88	15.11	15.09	14.88	14.98	15.03	15.27	15
Marketing margin of agent	54.36	55.14	52.54	52.89	55.62	51.44	54.12	56.08	74.49	73.52
Marketing margin of wholesaler	27.08	26.6	27.82	27	26.98	27.8	26.07	27.9	28.63	26.11
Marketing margin of retailer	82.71	82.17	82.92	82.14	82.66	82.92	82.35	83.3	82.73	82.29
Price paid by Consumer	1535.5	1547	1547	1545	1543	1534	1546	1546.5	1550	1543
Producer Share in Consumer's rupee	0.65	0.65	0.65	0.66	0.65	0.64	0.64	0.64	0.65	0.64

Table 3.14: Marketing Efficiency for Channel-I and Channel-II Respectively

Name of Village	*Marketing Efficiency (Channel-1)*	*Marketing Efficiency (Channel-II)*
Chandhara	4.93	4.27
Befin	4.28	4.28
Barsoo	4.62	4.34
Lethpora	4.52	4.39
Andrussu	4.50	4.25
Konibal	4.56	4.26
Dusoo	4.50	4.22
Kadlabal	4.55	4.20
Namlabal	4.15	3.96
Pampora	4.64	3.94

The value chain analysis describes the activities the organization performs and links them to the organizations competitive position. Value chain analysis describes the activities within and around an organization, and relates them to an analysis of the competitive strength of the organization. Therefore, it evaluates which value each particular activity adds to the organizations products or services and the level of arrangement of systems and systematic activates will make it possible to produce something for which customers are willing to pay a price. Porter argues that the ability to perform particular activities and to manage the linkages between these activities is a source of competitive advantage.

Porter distinguishes between primary activities and support activities. Primary activities are directly concerned with the creation or delivery of a product or service. They can be grouped into five main areas: inbound logistics, operations, outbound logistics, marketing and sales, and service. Each of these primary activities is linked to support activities which help to improve their effectiveness or efficiency. There are four main areas of support activities: procurement, technology development (including R&D), human resource management, and infrastructure (systems for planning, finance, quality, information management etc.). The term "Margin" implies that organizations realize a profit margin that depends on their ability to manage the linkages between all activities in the value chain. In other words, the organization is able to deliver a product/service for which the customer is willing to pay more than the sum of the costs of all activities in the value chain. Within the whole value system, there is only a certain value of profit margin available. This is the difference of the final price the customer pays and the sum of all costs incurred with the production and delivery of the product/service (*e.g.* raw material, energy etc.). It depends on the structure of the value system, how this margin spreads across the suppliers, producers, distributors, customers, and other elements of the value system. Each member of the system will use its market position and negotiating power to get a higher proportion of this margin. Nevertheless, members of a value system can cooperate to improve their efficiency and to reduce their costs in order to achieve a higher total margin to the benefit of all of them.

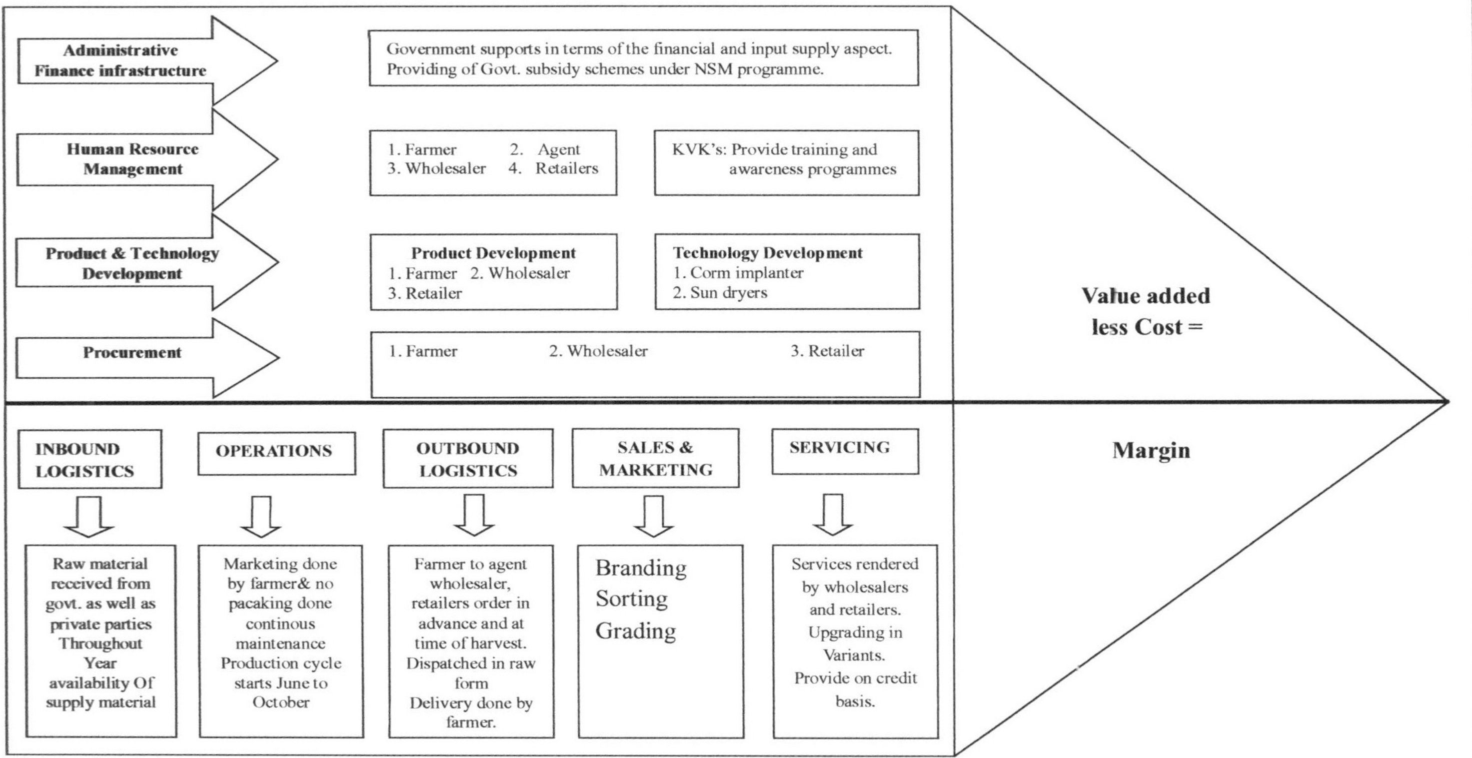

Flow Chart 3.5: Value Addition in Various Levels and Transactional Flow of Product.

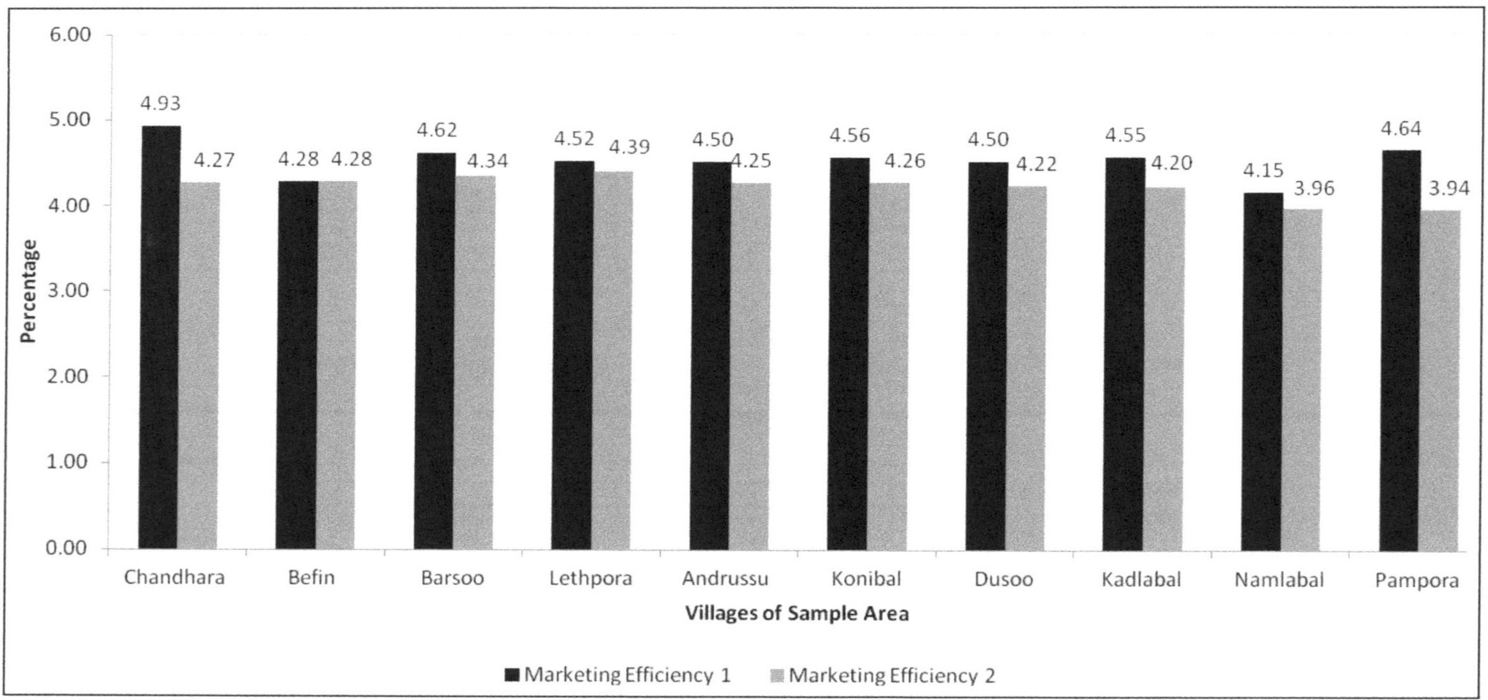

Figure 3.5: Marketing Efficiency of different Channels.

3.10 Value Addition

The Flow Chart 3.5 revealed that the in order to add a value to saffron spice chain, the product undergoes through nine activities viz: five primary and four support activities respectively. The primary activities comprises of technology development, Human resource management, Product and technology development and Procurement. While as Support activities comprises of inbound logistics, operations, outbound logistics, sales and marketing and servicing. The table depicted that in primary activities, for value addition process, the infrastructure development deals with the government support in terms of Schemes, finance and input sourcing to the producers for the saffron production purpose. While as the Human resource development deals with the aspect of (pre and post harvest) training practices for proper production method used by the producer for better saffron production as well as the awareness programmes conducted by the government departments for the welfare and benefit of the chain actors for smooth functioning of the saffron business.

The above cited flow chart revealed that the product and technology development deals with the process of production methods adopted by the producer and the using and adoption of new innovative and technological factors for better production purpose. In technological aspect, the use of corm implanters for implanting and sowing of the corms in the saffron fields and sun dryers for drying the raw moistured saffron spice properly in order to market the produce immediately during the high market demand for the spice for getting a fair price.

The support activity deals with the process of inbound logistics *i.e.* the procurement of the raw materials/inputs by the producer for the production process of saffron crop. Inbound logistics chain is from government to farmer as government acts as financing body and input supplier to the farmer for their production activity. The inbound logistics then leads to operational activity as it deals with the pre and post harvest practices and methods adopted by the farmers or producer for saffron production. Then next step delas with the activity of outbound sourcing *i.e.* the flow of product from the producers or farm level to the other chain actors for the value addition purpose. Usually the saffron crop after its production flows in two forms of channels viz: Channel-I and Channel-II.

The Channel-I comprises of Producer/Farmer → Wholesaler → Retailer → Consumer and Producer/Farmer → Commission agents → Wholesaler → Retailer → Consumer. These two channels are responsible for the value addition process of the saffron crop in the sample area respectively. The grading and branding of the saffron product for the marketing purpose comes under sales and marketing activity and marketing of the product deals with the servicing attribute associated with the product for satisfying the customers and consumers during the whole value chain process. During the whole value addition process that consists of these nine activities (primary and support) the total cost incurred by all the chain actors for doing the activity of value addition at each and every activity level is summated and deducted from the total value addition added by the chain actors individually at their respective stages. Hence the table depicted that from total total value addition, the cost incurred is deducted in order to get the margin of the chain actors at their respective levels. The Flow Chart 5.5 clearly revealed that the value addition linked with the product basically depends on the cost incurred by the various chain actors and their margin as well.

Chapter 4

Summary and Conclusion

This chapter explains the summary and conclusion derived from the present study on "Value chain analysis of Saffron in Kashmir Valley of J&K State" during the year 2014. Ten villages were selected on the basis of the highest area under saffron. The ultimate units, *i.e.* farmers, commissioning agents, retailers and wholesalers were selected randomly from each village so as to constitute a total sample size of 95 from the whole area under study. Collection of information regarding marketing of the saffron crop was done by visiting growers or saffron farmers, various markets of respective district as well as unorganized market outlets and contacting the different intermediaries involved in the marketing of the saffron crop.

The data collection was subject to analysis for examining the objectives of the investigation. The findings of the present investigation have been briefly summarized in this chapter.

4.1 Status and Importance of Saffron Crop

In terms of productivity, India appears to be at 5th position during the years 2012-13 and Pulwama district in J&K state depicted the higher scale in terms of area, production and productivity of saffron followed by other districts of the Kashmir region. Further, it has been identified that the price/kg of saffron appears to be highest the month of September and October due to the new arrival and harvesting season of the saffron crop. Mean while, the Compound annual growth rate for saffron in J&K. In terms of area area, production and productivity depicted a declining trend, whereas the study found that the farmer under saffron cultivation are residing in rural areas with an average landholding 5.5 to 8.0 kanals and contributing to the total production of 10MT.

4.2 Value Chain of Saffron in Sample Area

While determining the activities of chain actors in the sample area, the study concluded that there are primarily two prominent channels functioning in the saffron value chain. The channel-I comprises of farmers, wholesalers, retailers, government, and consumer and channel-II comprises of farmers, agents, wholesalers, retailers, government, and consumer respectively for the disposal and value addition of the saffron crop in the sample area. The government has an active role in field of providing the finance and relevant input sourcing for the cultivation of saffron purpose.

4.3 Economic Relationship between the Chain Actors

The study also determined the economic relationship between the stakeholders in both the channels of saffron value chain. It has been concluded that in the channel-I, the marketing cost incurred by the farmer was Rs. 2417.44 per 10 gram and physical loss of Rs. 448.76 per 10 gram. As far as the wholesaler was concerned, the marketing cost was Rs. 2549.60/10gm and for retailer, it was Rs. 2602.23/10gm for retailer. The physical loss for wholesaler was Rs. 267.17/10gm while as it was Rs. 150.27/10gm. The marketing margin of wholesaler was Rs. 138.23/10gm and for retailer, it was Rs. 1464.31/10gm.While study showed the relationship between the farmer, agent wholesaler, retailer and consumer with each other in economical terms comprising the channel-II. It was observed that the marketing cost incurred by farmer was Rs. 2417.44/10gm, while as the physical loss was to the extent of Rs. 448.76/10gm. As far as agent was concerned, the marketing cost was Rs. 2436.86/10gm and physical loss was having Rs. 67.44/10gm with margin Rs. 580.18/10gm. As far as the wholesaler was concerned, the marketing cost was Rs. 2549.60/10gm and for retailer, it was Rs. 2602.23/10gm. The physical loss for wholesaler was Rs. 267.17/10gm while as it was Rs. 150.27/10gm for retailer. The marketing margin of wholesaler was Rs. 138.23/10gm and for retailer, it was Rs. 1464.31/10gm.

4.4 Value Addition

The study analyzed two types of channels in saffron study area. The channel-I and channel-II were identified for the value addition process of the saffron crop. It was depicted that the channel-I adds more value as compared to the channel-II respectively.

4.5 Finance, Support from the Government, the Key Threats and Opportunities

The study revealed that the main sources of finance in the same area for saffron chain actors were Jammu and Kashmir Bank, Punjab National Bank, State Bank of India and Ellaqui Dehati Bank respectively. The main source of finance was in shape of Kissan Credit Cards and the Insurance benefits provided to the saffron farmers for their safe regards of saffron crop.

4.6 Share Created by the Chain (Price Spread Analysis)

It was found using Acharya's and Shepherd's Index techniques, that the maximum marketing margin were grabbed by the intermediaries (retailers, agents) followed by Wholesalers' leaving saffron growers an unfair margin. The empirical analysis depicted that in channel-I the Chandhara, Barsoo and Lethpora villages were more beneficial and fair for the saffron grower in terms of remunerative return for their produce in shape of producer's share in consumer's rupee. While as in channel-II, the producer's share in consumer's rupee was higher in Lethpora village followed by rest of the villages respectively.

It was found that the channel-I comprising farmer, wholesaler, retailer and consumer was more efficient than the channel-II and among villages, the chandhara seemed to be on higher efficiency side followed by rest of the villages respectively. While as in channel-II, the Lethpora village in sample study area was most efficient followed by rest of the villages respectively.

4.7 Conclusion

Saffron crop is highly remunerative and offers ample scope for employment generation the economic analysis in terms of costs and returns the crop is economic viable. There is much scope for making this crop more economically profitable provided sincere efforts are made by providing quality planting materials, introduction of sprinkler. Irrigation system, pests and disease resistance majors and efficient marketing system to expand the area as well as increase the production of saffron in the state. It was also concluded that in the sample study area, their exists two types of value chain channels viz; channel-I and channel-II respectively through which the saffron growers dispose off their produce for obtaining the return. Besides this value chain channels comprises of chain actors like; farmers, wholesalers, retailers and consumers supported by the government organization in channel-I and farmers, agents, wholesalers, retailers and consumers supported by the government organization in channel-II respectively. Thus can be concluded that value chain actors indulging in the processes of chain activities created a sought of value at each level in order to make a value added product to market by the nine (five primary and four support) activities. It was also noticed and concluded that the maximum marketing margins were earned by the intermediaries (retailers, wholesalers) who usually performed the task of grading, incurred the maximum marketing cost. The producer got a lesser share of the consumer's rupee in the channel involving saffron agents and vice versa. Most of the produce is however sold to agents, who usually visit the growers' places to collect their produce. The results also concluded that as the number of intermediaries reduced, the producers share in consumer's rupee and the marketing efficiency increased. It can thus be concluded that there is a possibility of increasing the producers share in consumer's rupee and marketing efficiency provided the number of intermediaries is reduced and government's intervention/major organized sector interventions in the marketing system is strengthened and encouraged.

4.8 Recommendations

4.8.1 Status and Importance of Saffron Crop

In order to increase the productivity of saffron crop in the J&K state, the new innovate and technological steps should be followed for the production purpose. The new areas should be taken under saffron cultivation in order to increase the production level. In order to overcome the menace of declining trend in area, production and productivity, the new technologies in terms of pre and post harvest practices should be adopted at earliest and installation of irrigation pumps and sprinkles should be installed in the saffron fields for the irrigation process of saffron land as it is totally rain fed. Providing proper timed high yielding varieties and adoption sun drying process. The price fluctuations have been the serious cause of concern to the growers, and there is a need to frame MSP (minimum support price) policy for this particular crop and/or bringing major organized sector players. It was also implicated that Farmers Interest Groups or Producers Groups with stakes in scientific cultivation, harvesting and drying/storage could be initiated.

4.8.2 Value Chain of Saffron in Sample Area

It was implicated to Strengthen and Support Value chain actors at each and every stage by providing quality inputs in terms of Seed, awareness as well as in terms of financial aspect. It was also suggested to incentivizing a modern marketing channel and developing a common brand name for saffron spice.

4.8.3 Economic Relationship between the Chain Actors

For economically sound relationship between the stakeholders, the fewer intermediaries involved chain is suggested and recommended for an economic viability of the saffron business among the value chains.

4.8.4 Value Addition

It was recommended to strengthen the value chain actors by active intervention and active role of the government in proper time at each and every stage of value added process in order to add a unique value to the product. The processing units should be set up in the area to promote the value addition process and generate local employment

4.8.5 Finance, Support from the Government, the Key Threats and Opportunities

It was suggested and recommended to establish other sources of financial support and other related schemes for the saffron growers and in whole for the saffron value chain actors for smooth functioning of the saffron value chain activities and The financial schemes available with government as well as Spices Board need to be popularized to enable the growers to avail benefits of these schemes. Extension agencies like KVK and agricultural department need to be strengthened and activated.While looking into the further enhancement and development of saffron crop for the economic upliftment of the state economy, the value chain actors and the policy makers should focus on issues related to expansion of trade to the international market, controlling the malpractices within the saffron trade business. In addition to this, there should be continuous focus in nontraditional areas with help of government support to create the better employment in the state.

4.8.6 Share Created by the Chain

It was concluded and recommended to adopt the fewer intermediaries' involved and direct channels for disposal and marketing of the saffron produce in order to increasing the producer's share in consumer's rupee. There is much scope to increase the producers share in consumer's rupee and marketing efficiency provided the no. of intermediaries is reduced and government's intervention in the marketing system is strengthened. In order to increase the efficiency level of the saffron channels, it was implicated that direct channel system model for disposal of the saffron should be adopted.

4.9 Policy Implications

The saffron production scenario in future has to been seen through competitiveness from other spice crops and measure has to be taken up to keep it profitable. The district having high but not up to mark productivity has to be focused strongly to improve the yield as the Compound annual growth rate for saffron in J&K in terms of area, production and productivity showed a declining trend. The technological adoption system is very poor and majority of the farmers in study area were low technology adopters. In terms of marketing aspect it is implicated that direct channel system model for disposal of the saffron should be adopted and incentivizing a modern marketing channel and developing a common brand name for saffron spice.. Furthermore, inclusion of other sources of financial support, insurance and crop schemes for the saffron growers & value chain actors for smooth functioning of the saffron business. The financial objective indicated that there is need of popularizing the government as well as Spices board of India schemes so to enable the growers to avail benefits. Extension agencies like KVK and agricultural department need to be strengthened and activated. The value chain actors and the policy makers should focus on issues related to expansion of trade to the international market, controlling the malpractices within the saffron trade business. If the saffron processing units should be set up in the study area, will promote the value addition process and generate local employment. The price fluctuations have been the serious cause of concern to the growers, and there is a need to frame MSP (minimum support price) policy for this particular crop and/or bringing major organized sector players. It was also implicated that Farmers Interest Groups or Producers Groups with stakes in scientific cultivation, harvesting and drying/storage could be initiated.

Bibliography

Acharya, S.S. and Aggarwal, N. L. 2001.*Agricultural Marketing in India*,Third edition oxford and IBH Publishing Company, New Delhi.

Ahangar, B G, Ganie H A and Padder U M. 2013, A study on institutional credit to agriculture sector in India, *internal journal of current research and academic review*, 1(4): 72-80.

Ahangar.B G, Ganie H A and Padder U M, 2013. A study on institutional credit to agriculture sector in India, *international journal for current research and academic review*.1 (4): 72-80.

Anchal Deepak and Sharma, V.K. (2009). "Price spread of litchi in Punjab". *Indian Journal of Agricultural Marketing*. 23 (2): 147-153.

Bandyopadhyay, A. 1989. Growth and instability in the production of main cereal crops of west of West Bengal and Punjab-Haryana, 1950-51 to 1984-85: a note. *Indian Journal of Agricultural Economics*, 44(2):158-168.

Bhagat, L.N. 1982, "Growth Rates of Output of Coarse Grains and Pulses in a Backward Economy: A Study of Temporal Spatial Variation in Chotanagpur", *Indian Journal of Agricultural Economics,* Vol. 37, No. 3, July-September, 1982, 392-399.

Bhat, G. M. 1986 "Marketing Cost, Marketing Margins And Price Spread Of Apple In Jammu And Kashmir through Different Marketing Channels." *Indian Journal Of Agricultural Economics*, 12(4).

Campion, A. (2006) 'Agricultural value chain finance in Peru', presentation at the Latin American Conference. 232-236.

Damodar R. N. 1996. *"Impact of HYV on Agricultural growth in Mahaboobnagar District"* Unpublished Ph.D. Thesis of Osmania University, Hyderabad.

Daniel A. S, William A. Matthews, J Medellín A and Adrienne B, 2014. The economic impacts of the california almond industry: a report prepared for the almond board of california 1-50.

Dantwala M.L, 1978. Future of Institutional Refonn and Technological Change in India's Agricultural Development, *Economic and Political Weekly*, 23-24.

Daroga S, 1981. "Imbalances in agricultural growth", *Indian Journal of Agricultural Economics*, 36(1): 71-75.

Dwivedi S, Sharma P and Bhat A, 2011 "An analytical study of capital formation in India: with special reference to Indian agriculture" *Economic Affairs*, 4(56): 359-363.

Gammage, S. 2009. *Gender and pro-poor value chain analysis.* Washington, DC, USA: USAID.

Gibbon, P., Bair J., and Ponte S, 2008. Governing Global Value Chains: An Introduction.*Economy and Society* 37(3): 315-338.

Gresta, F., Lombardo, G.M., Siracusa, L. and Ruberto, G. 2008. Saffron, an alternative crop for sustainable agricultural systems: A review. *Agronomy for Sustainable Development*, 28 (1): 95-112.

Grewal, S.S and Rangi, P.S, 1983 "An Analytical Study of Growth of Punjab Agriculture", *Indian Journal of Agricultural Economics*, 38(4), 509-519.

HAQ.I and SHAFI.S, 2014. Economic analysis of saffron cultivation in Kashmir valley of India, European academic research, 2(1): 1-2.

Herr, M. and Muzira, T. 2009. *Value chain development for decent work.* Geneva, Switzerland: ILO.

Iqbal, M. 1992. Marketing costs and margins of deciduous fruits in Balocistan. Outreach and transfer of fruit technology project in Balochistan paper 4, food and Agric. org. of UN and deptt of agric. and cooperatives Govt of Balochistan.

Iqbal, M. 1992. Marketing costs and margins of deciduous fruits in Balochistan. outreach and transfer of *Muhammad Sajjad et al., An investigation in to marketing channels* :484.

Iqbal, M. 1994. Marketing of Plums in Balochistan, Outreach and Transfer of Fruit Technology Project in Balochistan. Food and Agric. Org. of UN and Deptt. of Agric. And Cooperatives, Govt. of Balochistan.

Iqbal, M. 1995. Marketing of almonds in Balochistan.Outreach and Transfer of Fruit Technology Project in Balochistan.Food and Agric.Org. of UN and Deptt. of Agric. and Cooperatives, Govt. of Balochistan.

Jacques H. Trienekens, 2011. Agricultural Value Chains in Developing Countries: A Framework for Analysis, *International Food and Agribusiness Management Review Volume* 14(2).

Jha, B.K., 1994. "Growth and Instability in Agriculture Associated with New Agricultural Technology: District Level Evidence".

Kaloo J M, Patidar R and Choure T. 2014. Production and Productivity of Rice in Jammu and Kashmir: An Economic Analysis, *International Journal of Research* 1(4): 462-472.

Kamli, S. A. and M. H. Wani. 2004. Research Priority Setting in Saffron: *A Legendary Crop of Himalyas. Sher-i-* Kashmir Agricultural University of Science and Technology (SKAUST-K), Shalimar, Kashmir.

Kannan, K P (2011). Agricultural development in an emerging non-agrarian regional economy: Kerala's challenges. *Economic and Political Weekly*, 46 (9): 64-70.

Kaplinsky, Raphael and Morris, Mike, "A handbook for value chain research", IDRC. lessons from east African cut flower industries. *World Development* 37(2): 326-340.

Kasana, M. J. 2003. Distributive marketing margins of selected vegetables in tangi Tehsil of charsadda district. M.Sc (Hons) thesis, Deptt. Of Agric. Econ, Faculty of Rural Social sciences, NWFP Agric. Univ. Peshwar.

Kasana, M.J. 2003. Distributive Marketing Margins of Selected Vegetables in Tangi Tehsil of Charsadda District. M.Sc (Hons) thesis, Deptt. of Agric. Econ., Faculty of Rural Social Sci., NWFP Agric. Univ. Peshawar.

Khair S.M. 2000. Distributive Marketing Margins of Apple in Pashin. Msc (Hons), Deptt. of Agric. Econ, Faculty of Rural Social sciences, NWFP Agric. Univ. Peshwar.

Khair, S.M. 2000. Distributive Marketing Margins of Apple in Pashin. M.Sc (Hons) thesis, Deptt. Of Agric. Econ., Faculty of Rural Social Sci., NWFP Agric. Univ. Peshawar.

Kheirandish M and Gowda S.V. 2012. Marketing Efficiency and Price Spread for Saffron in Iran. *Trends in Agricultural Economics*, 5: 23-30.

Kumar M, 2011"Trends in Indian agricultural production in pre and post reform period" southern economist, 50 (15): 20-21.

Ladaniya, M. S., Vinod Wanjari and Mahalle, B. C. 2003. Price spread of Pomegranate. *Indian journal of Agricultural Economics*, 58: 800-811.

Ladaniya, M. S., Vinod Wanjari and Mahalle, B. C. 2005. Marketing of grapes and raisins and post harvest losses of fresh grapes in Maharashtra. *Indian Journal of Agricultural Research*, 39(3): 167-176.

Laven, A., Eerdewijk, A., Senders A. van Wees, C. and Snelder, R. 2009. Gender in value chains: Emerging lessons and questions. Arnhem, the Netherlands: *Agri-Pro Focus*. 1-18.

lawal J.O, Lawal O.O, Lawal R.T, Shittu and Muyiwa A, 2011. Profitability of value addition to cashew farming households in Nigeria, African *Crop Science Journal*, 1 (19): 49 - 54.

Lazzarini, S.L., Chaddad R. F. and Cook. L M. 2001. Integrating Supply Chain and Network. *Journal on Chain and Network Science* 1(1): 7-22.

Mali, B. K., Bhosale, S. S Shendage, P. N., Shindh, C. B. and Kale, P.V. 2004. Economics of production and marketing of Banana in jalagaon district of western Maharashtra. *Agricultural situation in india*, 60(11): 733-741.

Malik A Z. 2013. Assessment of apple production and marketing problems in Kashmir valley, Journal *of Economic and Social Development*, 9(1).

Malik A Z. 2014. Economics of apple cultivation "with special reference to south kashmir- india" Journal of Economics and Sustainable Development 5(9).

Malik, A.Z and choure. T. 2014, Economics of Apple Cultivation "With Special Reference to South Kashmir- *India "Journal of Economics and Sustainable Development*, (5): 2222-2855.

Mayoux, L. and Mackie, G. 2007. Making the strongest links. Geneva, Switzerland: International Labour Organization.

Mehdi S and Motiee, 2012. Saffron Export and agricultural value added in Iran. *European Journal of Experimental Biology*, 2 (1): 24-31.

Mehdi **S** and Reza M. Saffron Export and agricultural value added in Iran, *European Journal. Express. Bio. 2 (1), 24-31.*

Mondal, A. H. 1986. Marketing of pineapple in Meghalaya state. M.sc. thesis. Department of Economics and sociology. PAU, Ludhiana.

Munshi, A.M. 2002. Marketing and trade mechanism of Saffron in Jammu and Kashmir. Proc National seminar cum workshop on development of saffron in Kashmir. 83-88.

Nilakanth.R, 1980. "A note on agricultural production in India", *Indian Journal of Agricultural Economics*, 35(2): 89-93.

Nyoro, J. (2007). 'Financing agriculture: Historical perspective', presentation at the AFRACA Agribanks Forum.

Nehvi, F. A. and Wani, S. A. (2008). Saffron production of Jammu and Kashmir, Directorate of Extension Education SKUAST Kashmir, Shalimar.

Nehvi, F.A. Annual progress Report. (2007). On enhancing Production and Productivity of Saffron, The legandary crop of Jammu and Kashmir; Submitted to National Horticulture Board, 2007.

Pochanna, K., 2000. Growth and Instability of crop Production in A.P District-wise Analysis, Working paper series, Department of Economics.

Porter, M.E., 1990. *The Competitive Advantage of Nations.* Simon and Schuster.

Prabhavathi,Y. Krishna, N Kishore, T. and Seema. 2013, Analysis of supply chain of Spices in India: A case study of Red chilies, *International Journal of Scientific and Research Publications*, 1(3): 2250-3153.

Prabhavathi Y, Kishore K N and Seema, 2013. Analysis of supply chain of Spices in India: A case study of Red chillies, *International Journal of Scientific and Research Publications*, 3 (9).

Praduman Kumar and Surabhi Mittal, 2006. Agricultural Productivity Trends in India: Sustainability Issues, *Agricultural Economics Research Review,* (19): 71-88.

Puhazhendhi. V. and Jayaraman, B.1999. Rural credit delivery: performance and challenges before banks. *Economic and Political Weakly,* 16 (1): 175-182.

Pujari, A.G. (1998). "Marketing of Pomegranate and Ber". *Indian Journal of Agricultural Marketing,* 12 (1 and 2), 121-127.

Rabindra K R 2011. Agricultural Development in Assam Yojana (55): 34-36.

Radhika Rani Ch., M.Vanaja, Santosh.K.Bail, "Climate change and rainfed agriculture: rural development perspectives", *Journal of Rural Development,* 30(4), 411-419.

Ray, S.K. 1983. "An empirical investigation of nature and causes for growth and instability in Indian agriculture", *Indian Journal of Agricultural Economics,* 384: 455-459.

Ray, S.K., 1991. "Instability in Indian agriculture revisited", recent advances in agricultural statistics research, Wiley Eastern, New Delhi, 8-11.

Reerink, A. 2010. Mainstreaming gender analysis in value chain analysis and development. Women's Entrepreneurship Development and Gender Equality–East Asia.

Riisgaard, L. 2009. Global Value Chains, Labor organization and private social standards.

Riisgaard, L., Fibla, A. and Ponte, S. 2010. Gender and value chain development. Copenhagen, Denmark: Danish Institute for International Studies.

Rubin, D., Manfre, C. and Barrett, K. 2009. Promoting gender equitable opportunities in agricultural value chains. Washington, DC, USA: USAID.

Sajjad M, Khan M, Khan D, Mulk S and Nazir M, 2008. An investigation into marketing channels and margins of rice in district malakand 24 (3): 480-483.

Saraswat, S.P.; Dahiya, P.S. and Singh, P. (2006). "Production of peach fruit in Rajgarh area of district Sirmour in Himachal Pradesh". *Indian Journal of Agriculture Marketing.* 20 (2): 81-85.

Sarhad J. An investigation into marketing channels and margins of rice in district malakand, *Agric.(24).*

Sharan, S. P and Singh, V. K. 2002. Marketing of Kinnow in Rajasthan, *Agricultural Marketing.* 45: 2-4.

Shwedel, K. (2007). 'Value chain financing: a strategy for an orderly, competitive, integrated market. 56-59.

Sikka, B. K., Singh, Ranveer and Kumar, Rakesh. 1992. Profitability of apple cultivation in Himachal Pradesh, *Agricultural Situation in India,* 47(8): 637-640.

Singh, B., Prasad, V. and Yadav, S.R. (1997). "Cost structure and economic potentials of horticultural crops in district Farrukhabad, Uttar Pradesh". *Indian Journal of Agricultural Economics.* 52 (3): 631-632.

Singh, J., Chandra, A. and Kumar, A. (1997). "Prospects of fruits cultivation in canal command area of Bikaner, Rajasthan".*Indian Journal of Agricultural Economics,* 52 (3-4) : 627.

Singh, R. and Sikka, B. K. 1992. Production and marketing system of apple in tribal area of Himachal Pradesh. *Agricultural Situation in India,* 47(8): 196-200.

Tantry, M.A. 2014. Economics of saffron in Jammu And Kashmir State. *Review of Research,* 3(12): 1-11.

Thakur, D. Sanjay, S. Thakur, R.D and Sharma, D.K. 1994. Economics of off-season rice production and marketing in hills. *Indian Journal of Agriculture Marketing.* Himachal Pradesh Krishi Vishavavidyalaya Regional research Station, Kullu Manali, India. (8): 72-82.

Tomer, B.S Singh, S.P and Khadhar, R.K.1997. Marketing of grapes and citrus fruits in Haryana, *Indian journal of Agricultural Economics,* 52:642.

Tripathi, R.S. and Shama, M.L. (1999). "Marketing system of Apple fruit in Uttar Pradesh Hills". *Indian Journal of Agricultural Marketing, Conference Special.* 13 (2) : 47.

Vitonde, A. K and Bhargawa, P. N. 1991. Marketing of mandarin orange in Nagpur district. *Indian journal of Agricultural Marketing,* 5(2): 212-215.

Wani, M.H., N.A. Mir and A.R. Bhat, (1994). "Economic Viability of Apple Orchards in Kashmir", *Agricultural Situation in India,* December, 659-662.

www.ingramcontent.com/pod-product-compliance
Ingram Content Group UK Ltd.
Pitfield, Milton Keynes, MK11 3LW, UK
UKHW021956270726
14060UKWH00002B/544